AF885154

# The Small Perspective

Jack Nesigt

ISBN 978-0-692-97222-9

Editing provided by MJ

Dedicated to Her.

From a small active duty C-130 unit, a crew was selected to fly combat missions over Afghanistan through the course of a multi-month deployment. The crew was comprised of three commissioned officers: Major John Hammond, 1st Lt. Aiden Sheppard, 1st Lt. Daniel Carter, two non-commissioned officers: SSgt Sam McKay, SSgt Jonas Telford; and an airman: SrA Robert Ford. For all but one, this was their first deployment to Afghanistan. This is their story; a small perspective of a war that has endured for over a decade and that is all but forgotten.

For operational security, all locations within Afghanistan are given nicknames. In addition, most hotels, restaurants and other public places at which the crews stayed are also given nicknames. Photographs have been censored, as well as correspondence. Callsigns and names were changed from the real ones. As a result, the callsigns to identify the crew at the center of this story will vary and may be SHANCS 55 or MEDEVAC 55.
These were not their real callsigns, which in reality changed with every mission.

Events are described from the perspective of one of the crewmembers, and are in no way endorsed by the US Government, Department of Defense and/or the Department of the Air Force.

## Note about the author:

Jack Nesigt (nom-de-plume) is a former C-130 aircrew member and combat veteran who has deployed multiple times to various areas of the world, including the Middle East. The recipient of multiple Air Medals and other decorations, this aircrew member's recap of only a few events in Afghanistan, all taking place over a decade after the start of the war presents a glimpse into the challenges and successes members of our military face regardless of their responsibilities. He is also the author of a sequel to this story, The Larger Perspective, revealing how the decisions on where, when and who to send into the different Landing and Drop Zones of the world are made at the allied headquarters. Following his time in C-130s, Jack Nesigt was selected and assigned to a Special Operations unit in Florida, after which he resigned from his flying duties to transfer into America's space program.

Email him anytime at jack.nesigt@gmail.com.

Errors found in his books, including this one, will earn your name, or the name of your choice, an appearance in future writings, either as a character or a cameo appearance.

Unless otherwise noted, all photographs, including the front-cover and back-cover, are entirely owned by the author.

The author was a member of the crew, but to preserve anonymity, does not directly identify as one of the crewmembers.

# Prologue

## About the C-130H Hercules

Let's be honest: the C-130 doesn't look like the hunk demi-god Hercules after which it's named. It's a fat propeller airplane that has a nose that makes it look like a baby elephant. It's got a tail so big it looks like the sail on the back of a dimetrodon dinosaur (nice job trying to say dimetrodon). And it's old. Some are as old as our parents! That felt great. Here: go deliver this stuff to someone in a desert thousands of miles away, against bad weather and people shooting at you, using your dad's 1955 Chevy.

To us who flew it, though, it was a Cadillac. Every dent on those aircraft carried a story. This fleet of rag-tag propeller airplanes had proven that you could throw anything at them, from Vietnam to Afghanistan, and they would still take you home. Trust me, it's not like we didn't try to kill ourselves more than once with a C-130 through our own stupid actions.

The C-130 Hercules is a plane that keeps props cool in a jet age. This historical aircraft continues to be the longest continuously-produced military aircraft in planetary history; because it works. Lockheed Martin, the aircraft manufacturer, certifies that "there is quite literally a Hercules airborne somewhere in the world every minute of every day." And there is no mission except for spaceflight and air-to-air combat that a C-130 cannot and has not performed; from landing on an aircraft carrier or an ice-landing zone in Antarctica, to launching missiles against targets and refueling other aircraft, or airdropping medical supplies and performing electronic warfare, or conducting command and control missions, the C-130 is arguably the most versatile and flexible aircraft ever built.

The Herc has seen competitors rise and get crushed by its awesomeness, outlasting several airlifter designs. Used by every Major Command of the United States Air Force, and even by the Marine Corps (you know it's pretty awesome if even the Marines want to use it), the C-130 continues to be a workhorse for the American war effort.

## About the crewmembers

Sheppard was recently married. He missed his wife dearly. Which was surprising for someone of his stature and energy. A strong rugby player from the Air Force Academy who looked like he could bench-press a fully loaded C-130, his soft heart was evident to all in the unit as soon as he spoke. He emanated fun and positivity almost everywhere he went. Calling himself a 'meathead', the young officer was a gifted, intelligent asset to the military – and many knew he had a bright future ahead. The level of knowledge and understanding he had about flight operations were unequaled, surprising most of his superior officers. And even with all these bright accolades, he was an officer familiar with the word humility.

Leaving at home a girlfriend of 7 years, Carter was the other young officer on the crew. The one thing he looked forward to the most, both when returning from a mission or upon a return to the United States, was to see his girlfriend, either through video chat or in each other's arms. Unlike Sheppard, Carter considered himself part of the 'fitness protection program' and didn't particularly enjoy working out. He was more of a recluse, preferring to watch movies or read books to going to parties or talking much. Like Sheppard though, Carter had an intellect that impressed his superior officers, earning him consideration for various elite schools post-deployment, such as Test Pilot School. While he had a keen intellect and a strategic mindset, a normally restrained and supremely un-opinionated enlisted member would recall that Carter was "a badass officer and a smart mother-****".

McKay was an energetic crewmember; and this wasn't just due to his personality. He consumed enough energy drinks to keep the producers in business. The young man could take most jokes the crew threw at him, and they definitely took advantage of that. His patience with the crew was paramount, and he rarely complained, always instead focusing on the job and refusing to cut any corners – which kept us safe. What he did do was ask many questions about the missions, doing his best to keep the crew honest as to their flight operations. Throughout their combat flights, the NCO's most pressing concern was when would he get to ride his motorcycle again across the highways of Wyoming, wearing his leather vest emblazoned with the logos of a bike-club he was a member of? He was good-natured, almost always wearing smile and ready to laugh – a true asset to the team.

Ford was the youngest crew member. The young aviator was a Guardsman from another state who had signed on for 2 years of active duty and ended up with our joint unit. Also an avid motorcycle rider, Ford would often regale the crew with his dreams of buying a shiny sports car. The money he would earn from the deployment would go towards funding this new indulgence, while the time during the deployment would be spent searching for the dream car. A airman with pride, the young cadet of the crew was also deeply involved with the newest social apps, spending hours connected back to the United States through Snapchat, Facebook, Twitter etc.

Looking forward to marrying his fiancée, Telford was the mentor to Ford. His ability to switch from fun to work was impressive, allowing his positive demeanor to carry the crew through lulls while also re-focusing when complex work needed to be done. Normally hard to agitate, he always rolled with the changes and normally infuriating decisions with a composure that helped the rest of the crew adapt. As a non-commissioned officer in the enlisted corps, he also helped shape the officers above him, sometimes with the stern comments and reminders they deserved. He always took the time to mentor his junior loadmaster as well, often taking on the role of an elder brother to make sure life could continue. Never one to shy away from reason, his only goal was to get the job done so he could get home without problems to his fiancée.

Hammond was the highest ranking officer of the crew (and the eldest) while remaining the most approachable. His wisdom and tranquil attitude to the most complex missions we faced were traits that helped guide our rag-tag team through long months in the warzone. Through the toughest of times, it remains an assiduous challenge to remember any moment he lost patience or raised his voice. He spoke very fondly of his wife, patiently waiting for him as she studied for a PhD. He was more concerned about the wellbeing of those around him than his own and all he desired was being able to spend more time with his wife. His simple tastes were only compromised by his desire to buy an old 1980s Porsche, a dream of his since he was a child. The commander and deputy commander of the squadron certainly recognized his modesty and he was treated as an equal when complex events were considered, be they operational or events impacting our people. Never bragging or even mentioning his previous and impressive combat experience, his peaceful acumen and clear knowledge of operations were what allowed our team to succeed. The simplicity of this statement should not betray the depth of its significance.

## Table of Contents

# The Crossing

## First Leg – Departure Chaos

Deployment Day. Well, really, it was still night. A few minutes prior to 5 AM, deployers and their friends and families were gathering inside of the Air National Guard cafeteria known as the DFAC (Dining FACility) on the Guard corner of the civilian Regional Airport. It was a quiet and solemn time. Months of preparations, training and emotions reached their pinnacle on this day. Families and friends didn't want to talk about how long the deployers would be gone for, nor where they were going. They had already dealt with that news when they were first notified of their selection for the deployment. But this was it – the reality that we were leaving could not be ignored. Looking around, you could tell that for most spouses, parents and siblings, there wasn't much that could cheer them up right now.

"Why is it so early?" groaned a little girl who was sitting on her father's lap. His camouflage uniform was in stark contrast to the baby-blue dress his daughter was wearing – a contrast that was going to be ever more evident in the crew's daily lives through the coming months. The airman explained that the reason we were leaving so early was because we had to land before nighttime at their destination so as not to wake anyone up after that.

At 5 AM, the commander of the unit gave a few brief words of welcome and support before turning it over to the state governor. The governor had made it out early in the morning to wish us well. Not many governors bother to send off troops (it's such a common event now), so for some of us this was a special occasion, regardless of your political affiliation. Even though we were all members of the active duty force, we worked closely with members of the Air National Guard of his state, and the Governor wanted to show his gratitude. Even if it wasn't sincere, at least he had to pay for it with an early wake-up, so we'll call it even. A few more words from the Commander, and within ten minutes, it was time.

Time to leave. Time to say goodbyes and part ways. With today's technologies, we all knew we would be able to email, call using cell phones or calling cards – keep in touch. Even video chat, if the internet connection was good enough. But to part ways now was still arduous. A deep, empty feeling forming in our stomachs; the amalgam of yearning for those we were leaving and the exhilaration of going to America's most durable battlefield. Families were not concerned about missing their loved ones – they were *worried* about them. After all, we were departing for Afghanistan. Most of us, the men and women leaving for the warzone, were confident (maybe naively) in our competence and laughed off most of the threats. The family members had seen enough dramatic CNN / FOX / MSNBC reports to think differently. So passionate hugs and kisses were exchanged, while volunteers from the unit who were not deploying got busy doing a lot of the heavy lifting (literally and figuratively). To ease the burden on the departing troops, the volunteers drove different loads of luggage to the airplanes and ensured the airplanes were prepared properly for the flight.

After the whirlwind of emotions; arriving tired at the airport, a few short speeches and the quick, difficult goodbyes - all in about 30 minutes, it was time to start boarding the planes, or go to the Mission Planning Room and make sure the plans were complete by double checking everything, from routing, fuel plans, and permissions to land at different airports.

The contingent of 100 or so people leaving was considered and called a "normal package" from our unit: 2 planes, and 3 crews to fly them. In addition to the 3 crews came quite a few maintainers who were the sole reason why the planes were capable of flying (from engine mechanics to avionics specialists) and some extra personnel that made sure the crew's flight equipment functioned properly (from parachutes to oxygen masks). All these energetic and overall young troops would need supervision, so we can't forget to count a few supervisors among the contingent.

We would all fly these two planes to the desert. The three flight crews were going to share the job of flying the planes to their destinations, taking turns to rest over the next few days, and the maintainers were going to ensure they were in working order every step around the world. Because this was going to take a few steps.

The two planes had callsigns, and that's how everyone kept track of who was going to be where. GRASP 01 and GRASP 02 were each headed to different bases to pick up additional crews; GRASP 02 plane was to pick up fellow deployers in Little Rock, AR, and GRASP 01 in Atlanta, GA. Originally, the crew focused on this story was broken up for the flights over to Afghanistan. Sheppard, McKay and Carter were going to be flying with the unit Commander and the Chief of Tactics on GRASP 01, while Hammond, Telford and Ford were crewmembers on the other plane that was less leadership heavy, GRASP 02.

To understand some of the stresses associated with this crew mix, it's important to understand it from the perspective of the crew. The Commander, the person responsible for your next assignments and your future in this uncertain Air Force, was going to have the entire trip over to Afghanistan to observe you. And you are a brand new, first time deployer. In fact, you have never even crossed the ocean in this large propeller airplane before. The extra crewmembers (those that formed the third crew tasked with flying the two planes) were going to be resting on the other plane. That meant you were on your own; you were going solo with the Commander watching you on your way to the combat zone where there was no doubt he wanted to be able to trust your skills. This was going to be a test the entire way over (even if it wasn't officially). This was not a critique nor a negative aspect of our Commander; no one can be expected not to observe. It was simply the added stressor of having your Commander aboard. As the new officer, you were not going to put yourself in a situation where the Commander would doubt your capabilities through his own observations. The gap between a Lieutenant and a Lieutenant Colonel is more than just an added word: it's on average almost 15 years of experience. But when that Lieutenant Colonel is also your Commander, there's very little wiggle room to try and bond with the man. It would be considered inappropriate at best – so professionalism was critical for the voyage.

After a few hours, the Sun risen and most families gone home or to work, the plans were checked and printed and the airplanes were ready to go – everything was set. Both aircraft started their engines and completed their usual rituals to get to the runways. Per the plan, GRASP 02 took-off ten minutes before the other plane to go pick up their fellow deployers.

A couple of minutes after engines start, a light illuminated on the panel inside GRASP 01; the second plane that carried the Commander. The light signaled that there was a problem with the oil temperature in one of the engines. McKay stated that we had about 30 minutes before we would have to shut it down in flight, but the pilots and McKay agreed it was probably just a sensor that was overheating, as was normal on the ground. What often happened on the ground was the wind pushed hot gases back into the engine intakes. Most of the time, the light went away after take-off, so most of us were still optimistic, and the crew continued through our checklists and preparations for the flight. Regardless, this trip was starting off great!

We continued to taxi to the end of the runway and were given clearance to take-off. The throttles were pushed beyond idle and up to their max take-off power. And then the brakes were released. With a slight lurch, the plane started its race down the runway, gaining speed and slowly lifting off. In a slow, graceful left turn, the C-130 retracted its landing gear and climbing into the cool morning air, aiming for the rising Sun and leaving behind home for the next few months.

As the plane continued to rise through fluffy white clouds, the light remained stubbornly on, and it had been tauntingly glowing at the crew for 20 minutes already. McKay continued to observe the temperature on separate systems, but there was no sign of improvement, even after he tried to bypass different systems. Some fluids in the engine were simply too hot. Hopefully the next ten minutes would cool them enough that the flight to Atlanta could continue. We continued our climb-out and started to head direct toward our destination, passing through 10,000 feet of altitude.

Ten more minutes passed. Unfortunately, the temperature was just not dropping. The regulations were clear about what do in this situation, so the crew executed their checklists.

CP: Confirm number three? the copilot said pointing to one of the engine control levers in the middle of the control panel.

E: Confirmed.

CP: Number 3, coming down.

And he pulled the lever. With that, the engine was shut down in flight to make sure the problem wouldn't worsen. We now had only three good working engines out of the four used for flight. This sealed the fate of the airplane for the day and is considered an emergency in the aviation world (logical, since only ¾ of your engines are working).

We turned around and headed back to Clearwater, where we took off from, with three engines left. Carter got on the radios to notify both Clearwater and GLIDE 02 (the plane that had left ten minutes earlier) that GLIDE 01 was declaring an emergency and returning to Clearwater. As we had discussed before our departure, the other plane performed what is called a sympathetic abort and returned to Clearwater as well. For now, our contributions to the war effort were diminishing rapidly.

The early morning show-time, emotional goodbyes and nervous butterflies everyone felt as they left for the airfields of the battlezone ended up being just a false start. Twenty five minutes after take-off, both planes were back home, parked where they had been just earlier. Maintenance crews rushed to look at the engine that had issues. The news was grim: the engine would most likely have to be replaced entirely, which would take hours.

The leadership team got together to discuss their options. While they did that, everyone else gathered in the auditorium to wait. Some called their families excitedly to get a few more words in before we left, while most just did what they had planned to do in the back of the plane: sleep. We had, after all, woken up early from an already fitful night and therefore were still sleep deprived.

The contingency of an aircraft having to abort because of a malfunction had been discussed and a crew had been selected even before the day of the deployment. However, the verdict coming down from the Commander was different than what had been planned and expected. The members of our crew, the ones who were supposed to be reunited in Afghanistan after the crossing, would instead be reunited into one crew and left behind with a few maintainers. The original backup crew would leave on the functioning airplane with everyone else. They would fly to Atlanta and Little Rock to pick everyone else up before finishing the day in Newfoundland. The second airplane just had to get fixed and try to catch up.

With the decision made, everyone except for the few maintainers and the six basic crewmembers rushed back to the working airplane, found their spot, hung their hammocks, started engines and left.

What they left behind was our crew. We were the most inexperienced crew assembled among the three. Sheppard, Carter and Ford had never deployed before, and were still very new to the Air Force in general. It didn't get much younger than them. McKay had never deployed to Afghanistan before, only to other sections of the Middle East – completely different environments. Telford had very limited Afghanistan experience, having forward deployed there for a week at a time a few times while deploying to another section of the Middle East, as well. Hammond had already deployed to Afghanistan for several months before, and knew exactly what we were headed for. But Hammond wasn't from the unit, and no one had ever met him until this morning when everyone would deploy. No one knew who he was, and vice versa. That meant that two thirds of our crew had almost zero combat experience, and the highest ranking officer onboard had never met any of them.

The transit over to the desert was to be the "get-to-know-each-other" time-period for this crew to learn to work with each other. There was to be at all times a full crew that could be mixed together between the two planes, which meant that an adequate level of experience would be available for a safe crossing. Curiously, the leadership had decided to leave behind the crew that was the least familiar with each other and the least experienced. All experienced aviators, except for Hammond, were going to go together on one plane a day ahead. It had been suggested that the reason why this happened was because there had been no other crew that could be reliably trusted, but this may just have been hot air to boost our confidence.

Because the repairs were going to take so long, our crew would not have enough duty day to fly to our destination. Crews are limited by regulations from flying more than 12, 16 or 18 hours depending on the type of flying or number of crewmembers. We were limited to 16 hours, and our destination was about 12 hours away when you included a quick fuel stop. But we had already eaten through about five hours from when we first showed up at the Dining Facility to now. Which meant that if we were to take off immediately, we would still go beyond our allowable time to fly (5 hours of work time already accomplished + 12 hours of flight ahead = 17 hours). It was decided that we were going to leave the next day, exactly 24 hours from our original departure time to keep flight plans and authorizations easy to change.

Everyone went home or back to their hotels for one last, again, night in the United States. It was a bitter sweet moment.

The next day we got to the airplane, this time with no emotional goodbyes or governor visits or speeches. This was a largely invisible departure, and our crew was on its own to get the papers and plans together. Some members from the unit that weren't deploying were surprised to see us still, thinking everyone had already left the day before. The flight plan was different from the day before since there was no longer any need to pick people up in Little Rock and Atlanta; the plane yesterday had taken care of that before moving on to the first destination: Newfoundland, Canada.

Because we were now a basic crew complement, our crew was divided into the "front-end" crew, those members that would spend the flight on the flight deck, and the "back-end" crew – the two loadmasters who were masters of the cargo bay compartment. The line that divides the two crews is called "two-forty-five", a partition between the upstairs flightdeck and the cargo compartment. The front-end crew received their flight orders and took a quick look at the weather and route of flight, while the back-enders proceeded back to their repaired airplane and ensured everything was reloaded and rigged properly for flight.

Before we took off, the crew got together for a final, informal brief in the back of the airplane:

Hammond: Ok. Hey guys, I'm John Hammond, I'll be in the left seat today. Purpose of the mission is to get us going to Afghanistan. We're Alpha 1 and we've got... 42 K on the fuel. Weather is as you see it outside and we're not expecting anything throughout the mission today. Intel update we already got inside. Let's see, we're... 18 onboard? So we're looking for 18 in case of a ground egress. In case of an emergency, the Crew Entrance Door will be the primary exit, but don't pass up a good exit if you see one. Loads: chocks if you can, and Daniel grab a first aid kit on the way out if you can. We do have flares aboard so we'll meet 600 feet off the nose of the airplane. If you get separated from us, give us a touchdown signal to tell us you're ok, otherwise we'll assume something's wrong and make our way to you. I'll make the call on whether or not to send someone back into a burning airplane to get someone. Daniel will make the call if I'm the one stuck.

Otherwise, make sure your scarves and rings are removed, are there are any questions or reattacks?

The crew had nothing more to say, so we all got to our seats and started the engines. Half an hour later, we took off and flew to Minneapolis, which was to be the fuel stop to make sure we could make it to Newfoundland. The plan to get us to our final destination for the day wasn't even complete yet, but Tanker Airlift Control Center (TACC), the controlling agency that created the plans for this flight suggested that we take-off, and when we landed in Minneapolis the rest of the flight plan would be complete for the final leg to Newfoundland. We would just pick up the rest of the plan at the Minneapolis Command Post.

Minneapolis wasn't that far: two hours flight time. But an hour into the flight, Carter realized he forgot his flight helmet. Being in flight, he stressed out, wondering if he would even be allowed to fly when we got to Afghanistan. The flight helmet was a mandatory tool needed for night missions involving night vision goggles, and it was also mandatory for combat landing zones. Unfortunately for Carter, turning around was out of the question: we were now committed, so we had to reach another solution.

The plane landed uneventfully at Minneapolis International Airport and parked on the Guard ramp which was about to refuel us. Meanwhile, the officers went inside with McKay to receive the rest of the flight plan. While Hammond and Sheppard waited for the papers to print, Carter got on the phone and called Sarah, the Unit Deployment Manager.

Sarah had been responsible for all the deployers being ready for this mission, and it was a tall task: 100 or so people all had to have uniforms, training, gas masks, bags etc. And to corral hundreds of independent, mainly Type-A military members (particularly aviators and officers), was a very tough task at best. And Carter was reminding her that the job wasn't over even after they had departed. There was always someone with something. The support aspect that goes with preparing for war is often, regrettably, a forgotten facet of the war effort.

Fortunately, she reassured Carter and told him that the helmet would be shipped directly to Afghanistan by mail over different networks including military airlift, and until then, he could use a loaner already on site.

With those reassurances, Carter rejoined Hammond and Sheppard to look at the plan for the next flight portion of their day. We had 8 hours of flying ahead to get to Newfoundland. After taking a quick look at the charts, and the fuel plan, the crew walked back to the airplane and prepared to leave the busy international airport that was Minneapolis.

Once airborne and leveled off at our cruising altitude, our crew had the next eight hours to talk and get to know each other. And meeting older, more experienced military officers was always special, particularly at twenty-three thousand feet, with a view of the Great Lakes. Meeting military people always carried the risk of meeting a brash, proud and abrasive veteran. Arrogance was often a word that was ignored in some officers' minds. Humility was another concept often unknown to other officers. Some, though, would laugh and joke around. Some were passionate about guns and 'Merica, believed the President to be a Devil Cult leader, while others listened to National Public Radio and were passionate about finance. The diversity was frustrating – you could never simply expect someone to be a certain type of person. Major Hammond's most pressing issue was to find an old Porsche that had been his dream since being a child, and completing enough small deployments to avoid being selected for a 'super-deployment' which would take him away from his wife for one full year away. His soft-spoken demeanor, constant laughing and willingness to listen even to the most inexperienced crewmembers allowed our crew to bond quickly. The fears of having an overbearing, I'm-experienced-and-I'll-show-it superior officer were quickly dying down. We would be a crew, and not a man with his supporting cast. Hammond would, in fact, reveal himself throughout the rest of his deployment to be one of the nicest and most qualified officers the Air Force had to offer.

So for the next eight hours, our crew talked and got to know each other. As the plane crossed over state lines, thousands of feet over Chicago and small towns in Canada that bordered the Great Lakes, our crew was introducing itself to each other; Clearwater wasn't the original home for anyone on our crew, so we learned about how we had ended up in the same unit from various corners of the world. Our conversations were occasionally interrupted by Air

Traffic Controllers passing us on to their neighboring controllers, reminding us that we were still traveling East, away from home.

Eight hours later we approached Newfoundland. The airport we were aiming for was a legendary place within the Hercules community, having been a northerly staging airport for oceanic crossings for many years. The weather had rarely been anything better than rainy and cloudy throughout most of the year, and strong winds, snow and ice were what you had to expect. That evening was no different.

As we got closer and closer to the airport, Sheppard, who was in control of the approach, was about to perform the final turn to end up aiming at the runway. Unfortunately, we were flying through clouds, and we lacked visual cues to be sure when to turn, and the winds were very strong, making the turn the computer predicted inaccurate. So strong that when the plane performed its final turn, it had to be turned twenty seven degrees into the wind to keep going in the direction we wanted. This phenomenon where the airplane was flying sideways was called crabbing, and wasn't unusual. But it was still disconcerting when we were in the weather and running low on fuel. As we kept descending, there was nothing but clouds still. We could only go down to 200 feet above the ground, or else we'd have to try the approach again and hope that in between the approaches, the weather improved just a bit.

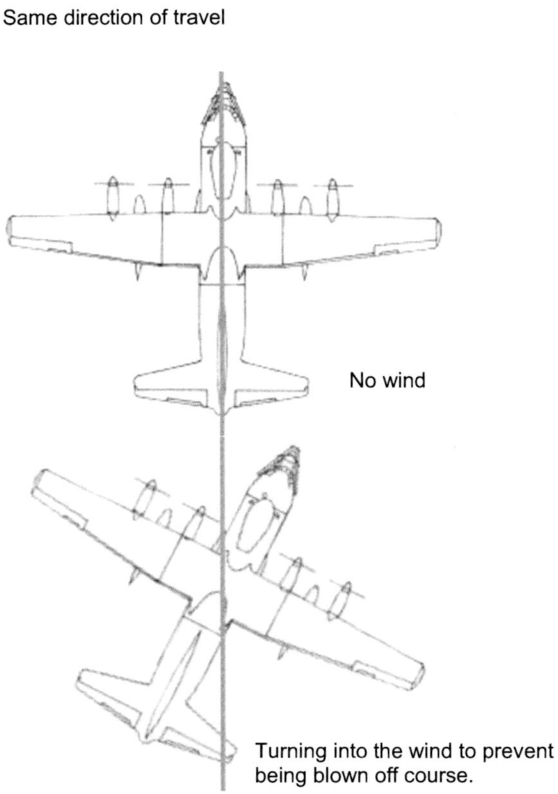


We fortunately ended up below the clouds around 300 feet above the ground and saw the runway, with our heads turned slightly to the left, the nose of the airplane aiming far to the right but the airplane still tracking toward the runway. Piercing below the clouds also revealed a new challenge in the night: rain. This meant that the runway was going to be wet, more slippery, and of course, the splattering drops were causing visibility issues on the cockpit windows.

Masterfully performed, the plane landed without any problems and we taxied to the parking spot in front of the small airport office that was going to host us. We had some issues with parking: the marshalling agent asked us to back further into our spot than we were comfortable with. But we did it anyways, and as the engines roared into reverse, the marshaller was blown with rain water and strong winds, making his life more miserable than he had probably anticipated.

After the engines shut down, McKay did his walk around and saw fluids coming out of one of the propellers, but decided not to panic this early. He decided to blame the rain for making things look worse than they were. He and the maintainers would look at it in the morning. Until then, the crews got their overnight bags and were welcomed aboard a large bus that offered free beers to all their guests. For many, this was finally the first sign that we were gone. Deploying, going to war and not going home any time soon – some of the guys were ready for that beer after a long flight over. We had just arrived in another country and were headed to our first hotel.

The front end crew quickly found out the departure time for our flight the next day. Our crew checked into the same hotel that most crews had stayed in for years, and after a half hour to freshen up, we met up to head into town, even though it was already between 9:30 PM and 10:00 PM.

Everything was within walking distance, and we went to a small pub in the old town that was a block away from the wharf where large cargo ships were docked. The quiet little town had the second largest per-capita income of Canada, and its industries were booming after a collapse of the fishing industry in the 1990s. It was still drizzling from the earlier rain, but large floodlights showed the relatively calm bay that the ships were docked in. The quaint town kept all its historical heritage, having been one of the first towns founded by colonists in 1583. The pub we chose was one of the few places left open and serving meals, which we all ordered (around 10:30 PM). After that, most of us turned in for the night, knowing that we were ocean crossing the next day.

## Second Leg – Over the Ocean

Whereas most airliners have no issues crossing the oceans today (in the mid-2010s), the C-130 has a few more limitations that make these crossings more interesting. The fuel load is much more limited already, and the navigation systems cannot always be trusted because they were designed for military operations. A perfect navigation system meant no one should touch the programmed software, but combat crews needed the ability to adjust. So the system couldn't be perfect for it to be useful for our warplane – we had to be able to go inside and tinker with it (and make it better...). Of course, if something were to happen while we were traversing the ocean, we would be far from any sort of help – radio communications are available, but still difficult due to the great distances. Airliners pass above at almost twice our speed, waving at us in pity, and if we had fuel problems, then that would be it. Hopefully we would have enough to make it to a divert airport or to turn around (although we may just want to just crash right there out of shame). The planning of these kinds of missions is very different, and a lot of contingencies are added into the plans.

Similar to most flights from North America to Europe, we were going to be flying overnight as a red-eye. This did mean that our take-off was in the evening, and we had to check out of the hotel by noon. That meant we had to find a place to rest and leave our luggage for 5 hours.

A few of us went to Subway to get a sandwich for the overnight flight. After takeoff, our first stop would Mildenhall Royal Air Base, in the United Kingdom, but we were just stopping there for fuel. The final destination was the capital of an Eastern European country (formerly Communist, but now a member of the European Union, NATO and in the future the Eurozone, showing how much our world can change), nicknamed Syldavia. That meant we were going to be accomplishing a 13 hour flight, so food for the trip was essential, as was caffeine. We purchased multiple bottles of Coke and Red Bull energy drinks, those who dipped got a few more rolls, and those who smoked were out of luck.

After bumming around the town for a few hours, the shuttles from the air terminal who was hosting us started to arrive. No beers this time, since we had to fly. The trip was a quiet one as the Sun was setting on the bay and port that had been home for less than a day.

Arriving at the terminal, the back-enders (loadmasters) went to prepare the cargo compartment for the long journey ahead, while the front-end crew began to plan the mission,. Large paper charts were unfurled on the table, and laptops were used to compute the flight over the ocean, taking into consideration the weight of the airplane, and how it was going to change as we burned fuel. This would also affect our fuel consumption, which over an area that had no gas-stations, was a slight concern for us. The winds and temperature were also considered when calculating time in flight. Airports close to the coastlines were mapped out in case we had to make emergency landings. Reporting points, to let ground controllers know where we were (since radar services didn't extend that far out over the ocean) were also charted. After going over the fuel calculations twice and two hours later, the crew was ready to print their paperwork. Maintenance crews and the Flight Engineer fueled the airplane almost to its maximum capacity, and the loadmasters were almost done with their own pre-flight checks. Everyone was ready.

It wasn't raining anymore, and McKay's concern of a leaky propeller were assuaged when he did his walk around the airplane: there was no puddle of fluids. But even in the cool night, the cloud cover still did not inspire confidence, and we were ready to take-off as soon as we could to try and leave behind the bad weather.

The loadmaster, like for every engine start before that, went outside onto the tarmac of the peaceful, civilian international airport and stood a dozen feet in front of the propeller while the flight deck crew initiated the engine start.

P: Starting engines checklist.

E: Clear number 3.

L: Number 3, clear.

P: Number 3, turning.

E: Light... drop... RPM... fuel flow... ignition... parallel... oil press... hydraulic press... starter... light out... series peak TIT 770.

E: Remove external equipment and fire-bottle. Clear number 4.

L: Number 4, clear.

P: Number 4, turning.

The crew went through the same process for the next three engine starts. For all engines, it was about the same thing. There had to be oil pressure indications by the time the RPMs reached 35%. At that same moment, fuel flow and ignition were both supposed to occur, by which time we should see the engine temperature skyrocket. The ENG START switch was released at 60% RPM. The 24 gauges on the dash were individually monitored for every engine start, making sure the starting engine was behaving as described in the aircraft manuals. The numbers were memorized by the crew over years of training to make sure they could catch the errors quickly and immediately. The RPM gauge had to start spinning, and after the fuel flow needle moved, the Turbine Inlet Temperature gauge had to also move immediately after; the fuel flowing had to be ignited, after all, for an engine to work properly.

The loadmaster outside didn't worry about the gauges. He looked at the actual engine and made sure the propeller was spinning as advertised, without spewing fluids or flames. The latter two were not part of the normal operations of an engine.

All four engines started without incident ahead of the long flight over the cold North Atlantic Ocean we were facing. The engineer continued his fast flipping of switches as he controlled everything in the aircraft systems; fuel was being diverted from one tank to another, electrical buses were coupled to different generators, and aircraft pressurization was tested, along with the hydraulic system, to make sure brakes would function properly.

E: Starting engines checklist, complete.

We performed a final check of the navigational systems before taxiing to the end of the runway. As soon as we got clearance, the engines were throttled up, and slowly the very, very heavy airplane started to roll down the runway. After gaining enough speed, it lifted off into the air, slowly but surely and immediately disappeared into the clouds.

On the radar screens, the crew could see a few thunderstorm cells in the middle of their flight path. Normally, that's not much of an issue: you simply go around it and avoid the storms. The problem with this flight is that fuel had been planned to the minute due to the limited availability. Avoiding a thunderstorm meant going off course and adding more time to the flight plan. Would we then have enough fuel?

Picking our way around thunderstorm cells for a tense hour, the crew was then able to head back to the original course line. Doing some quick calculations, the fuel load was still enough to get across the ocean and most of the British Isles to land at Mildenhall. As the plane continued to move forward, we would be burning less and less fuel to maintain our airspeed since we would be getting lighter and lighter. So far, so good, and the crew was now looking ahead to the long, dark expanse of empty ocean.

Ahead of us was nothing but pure blackness. No streetlights, no cities below to fly over, not even ships were visible because of the cloud deck. Radios were virtually silent because most airliners used a text messaging system to report their positions. Even our aircraft was silent. We were only supposed to check in every hour or so; there wasn't much to say between then.

Although we were still together as a crew and we had a few maintainers in the back as well, the feeling of loneliness was still omnipresent. Everyone in the back was lying down in hammocks or on litters trying to sleep a little. Everyone up front stayed fairly quiet as well. A few quick snippets of conversation would break up the quiet atmosphere, but otherwise, not much was said. It's unknown if it's because we were getting tired, if we were nervous with the Atlantic crossing and everything that could possibly go wrong, or if we were just individually lost in our own thoughts.

While the darkness and clouds below did not allow us to see the ocean, looking up brought us an entirely different sight: stars. Unlike anything we had ever seen before. There were no clouds above us, and no lights for hundreds of miles around, which meant no light pollution. Of course, regular pollution (smog) was absent as well, being far from any sort of large sites of civilization. The blackness around us was infinite. But above us, a blanket of small pin-sized lights covered the entire sky. The starlight was visible but faint, and it was everywhere - infinite. There were barely any spaces for darkness to spread from. As if someone had used a paint brush to continuously jerk starlight onto a black canvas. Some stars were slightly brighter than the rest, but most were simply a faint white, and all looked equal – a mural of scintillating lights. There was a never-ending number of them – for the first time in most of our lives, we could see the vast expanse of the galaxy that harbored us. For some reason, we had never seen anything like it even when flying with civilian airliners, but perhaps it was due to the angle the cockpit windows were at, or the very dim lighting on the flight deck, or a combination of both. Regardless, if at any point during our voyage we wanted a break from the darkness, all we needed to do was look up. We would see the glittering stars like a nightlight reminding us of our place in this world: fairly small and infinitesimal. A perfect message to gain perspective before heading to a battle zone...

The quiet flight continued for many hours. While it would sound boring, there were moments of excitement that would come up when the computer systems would display the message GPS NAV FAIL. The message meant we were running on internal sensors to navigate – no GPS satellite or other external system could help us. With winds that blew across us at 115 knots, loss of our navigation system meant we could be blown off course relatively quickly. Fortunately, the crew had plans and techniques in case this happened and to stay on course – using mathematical formulas and even archaic (by 21$^{st}$ century standards) rulers and dividers to plot chart positions, the aircraft's position was continuously monitored and a few degrees of heading changes were all that were needed to keep the plane on course. While that may not seem like much, a single degree of error would mean the plane would be off by 5 NM within an hour. So if it was more than 1 degree off, the plane could be dozens, if not more, of miles off by the end of the flight. That could be the difference between Southern England and Northern France. Luckily, after adjusting the different electrical systems and manipulating the sensors to regain reception, our systems would come back online and we would realize our calculations were accurate and our position didn't drift very much. Even through these moments (total of 3 for this crossing), the crew remained quiet except to discuss different options to help remedy the issues. After confirming our position using other systems, the silent moment of panic would pass. Attention would return to the vast ocean ahead, the miles counting down until we would hear once again the voice of a controller, and even further up, lights from the coastline.

But long hours still had to pass. We passed the Equal Time Point, which meant that now, regardless of what happened, it would take us less time to continue ahead to the British Isles than to turn around for Canada. We were halfway to the United Kingdom, but that didn't include the time it would take to get to Syldavia. The only comfort coming around that Equal Time Point was another set of calculations that were performed and confirmed that the fuel remaining was above what was needed to reach the Royal Air Force Base. We would make it with the fuel we had, which took away another stressor hours into the crossing.

A quick bathroom break to head to the back of the airplane revealed how lonely the crew up front was: everyone was sleeping, save for the few exceptions that were lying down, eyes half open to keep reading their Kindle e-books or finish their movies. This crossing was nothing to them: nothing to worry about, no windows to look through. The biggest concern on some of the maintainers' mind was what kind of night they were going to have when they got to Europe. Others silently missed the loved ones they left behind. Life was continuing without them.

After a few quiet and dark hours, we could finally start to hear radio calls coming from the Irish coast. We were getting closer. A look at our computers told us we were only a few dozen miles to their airspace entry point. We would quickly be back within controlled airspace, and soon after that, within sight of land. The cloud cover was no longer here, so we could start to see some ships, and more impressively other airplanes that were converging on the oceanic reporting point and the airspace entry point – an invented point in the sky that had nothing significant to look at visually, but that could still be seen thanks to planes coming from the West converging toward the point.

We terminated overwater procedures by the crew as soon as we received ground-based navigational aids, about a hundred miles from the coast. That meant that twenty minutes or so later we would be hitting land. And we did.

The number of ships started to increase exponentially, which was our first clue that we were approaching land. And we then saw a few villages that confirmed the presence of land, piercing the blackness of the night. It was too dark to see the actual coastline, particularly from our altitude, but we could sometimes tell where certain portions were supposed to be depending on the lit streets that linked two coastal villages. These strands of light connected villages with each other or simply faded into the dark countryside or terminate at a town, like a river of yellow light vanishing into a dark, black ocean.

As we started to fly over Ireland, we noticed major thoroughfares webbing across the country. We then flew over the light oasis that was Dublin, just as the Sun started to rise in front of us. The massive city, from on high looked like it was bustling with life, even though it was still close to 5:00 in the morning for the Irish down below. But the lights and the Sun rising signaled to us also the end of the crossing and a new day. Today, we would fly from the United Kingdom to Syldavia.

Another signal that parts of Europe were waking up, in addition to the lights, were the radios. As we came into range of the islands, we got to speak to controllers, but more importantly, we were back on a network that airliners were also a part of.

The occasional chatter but general quietness still reminded us it was dark. But as we crossed over Ireland toward the United Kingdom, more and more airplanes started to take off and join us in the orange skies of Europe. They were carrying the early-bird passengers going to their meetings somewhere else in Europe, and the pilots operating the airplanes joined in on the conversation. As the Sun rose, so did the number of people chatting on the radio, creating a wave of necessary conversations as the air got busier. Our crew consequently also regained some of its energy and focus on the flying, feeling part of this new day.

We crossed into British airspace immediately after passing Dublin, and a new cloud cover presented itself in unmistakable English fashion. A cloud cover which was in fact a fog that was even present on the ground in Mildenhall. We were still able to see far enough to land on the fogged runway. After we parked and shut down the engines, the feeling of a quiet early morning wake-up was strong for those who had slept in the back due to the fact that most of the base was inactive. We had landed just minutes after the required quiet hours, which meant that the base usually didn't start operating at full pace for the next few hours.

The front crew went to get their next leg's paperwork, and within an hour, we took off again with our fuel tanks topped off. We headed for the rest of the day to Syldavia. After ten hours walking around the same plane, we had gotten one tenth of that time to walk and breathe fresh air before we launched again for the cloudy skies over Europe. As we progressed eastward, we crossed over several different countries, hearing controllers speak to their national airplanes in their language, and every other foreigner was spoken to in accented English.

It was of course very busy radio chatter all the way to Syldavia, given the thousands of flights happening everyday across all major European cities, from Paris to Rome, Athens to London and Berlin to Madrid crisscrossing paths with ours. The beautiful landscapes of Europe that we could occasionally see through clouds were different from the United States. The vast expanses of Kansas did not exist in Europe and as autumn settled in Europe, patches of colorful red with some green holding out could be spotted.

Five hours later, we landed at the Sbrodj International Airport. Most of us got our overnight bags and waited by the shuttle outside the terminal for the few maintainers that were sealing the aircraft for the night. When that was done, we piled into a small shuttle and were whisked away to a luxury hotel in downtown Sbrodj.

Our long day that had started in Canada was ending in Eastern Europe, and the night ahead was going to be one of celebrating our successful crossing and our last night of absolute freedom before we headed into the deployed environment.

The hotel staff was well prepared and had most of our rooms ready, and check-in was relatively quick. Also, for the first time we ran into crewmembers of the first plane that had run a day ahead of us. They were leaving the next morning, but far earlier than us. As a result, we were all reunited for our first night out together, all as deployers. It was also, ironically, our last night in the Western World, where there would be reliable Internet, taxis, bars and a luxurious variety of restaurants. That's not to mention the last place we would be at where the police force wasn't military or where a majority of the people walking around would be wearing a diversity of clothes (aka, not uniforms).

We went into the city and discovered how far the capital of a former communist nation had come in developing itself to 'Western' styles of capitalism. The architecture resembled many of the Western European palaces such as those in Paris, and like in Paris or Berlin, new bars and small restaurants and bistros flanked closed but lit clothing shops with the latest styles. Of course, this was the first time we all went out in a country where we were visibly American. Canada had the benefit of speaking English with a similar accent, and most people dressed the same way. The few clues that indicated we were from out of town was the fact that we travelled in large groups, and of course, some were louder than others. Thankfully, most of the loud people from our unit that were deploying were on the sister ship, and they were all in bed.

But we were different now. Loud, poorly dressed compared to elegant European tastes and most (with some exceptions) of us very muscularly built, the signals that we were US military was evident. Understandably, we weren't necessarily always welcome. Even if it weren't for the military aspect, no restaurant wants 30 or so people barging in and taking up all their spots and time because it can scare away other clients. Fortunately, some of us understood this, and we split into smaller groups.

Finding a place to sit was a challenge, and ordering in a non-native tongue difficult as well. Some of our fellow aviators were from Texas and Alabama, and their thunderously loud voices confirmed it. Mix in a little bit of alcohol, and these folks would become louder, and start defending the honor of their individual states as far as their lungs could project. Usually around that time was the cue that we should head back to the hotel, and so we slowly started to wander back through the well-lit, lively streets. All around us were other bars and restaurants, patio dining and bistro seats open. Much of the neighborhood reminded us of Paris.

The city itself was preparing for a big football match, which we were told was going to be an incentive to be indoors and in our rooms once it was over. There were concerns about possible riots or violence if the match didn't end well for Syldavia (even if it did, concerns existed for out of control celebrations).

Even before the match began though, it was still strongly recommended we not travel alone outside the hotel. This wasn't due to the fact that Syldavians were particularly dangerous – the same rule applied to the UK and Canada for example, but there had been past experiences that motivated such decisions. As a result, keeping track of everyone while we were headed back to the hotel was a challenge. Some wanted to drink more and diverted into bars, while others were tired and rushing to the hotel, mumbling "good luck" under their breath and "I'm not with them" to passersby.

Once back at the hotel, most still enjoyed a few more drinks at the hotel bar where the match had begun playing on the bar-screens. After another hour or so, though, the crews started to head for bed, knowing full well we still had to fly the next day.

So the crews turned in, most of the rooms on the fourteenth or fifteenth floor of a five star hotel in a European capital. The city lights bleeding through the curtains and the cheers of the football fans lullabied the crewmembers to sleep. Most of the flyers fell asleep quickly after having been awake and flying for over a day, including crossing the Atlantic Ocean.

## Third Leg – Deployment Starts

We had been scheduled for a minimum turn time in Syldavia, which meant the next morning, we had enough time to wake-up, get some breakfast and then we'd have to take the bus to get to the airport.

A note about the breakfast will be made though. The hotel offered bite-sized pastries, along with fruit, cooked meats and eggs, all of which tasted absolutely amazing. This was the last breakfast that would not be served inside a military cafeteria, and it was enchanting. The buttery pastries tasted so wonderful, they easily could have been imported directly from France. Everyone ate for as long as they could before the final moment, when it came time to get on the bus and go.

Once the bus arrived at the international airport, the flying crew went into one of the airport's offices to gather charts and flight plans to see where we were headed, but more importantly, how we were going to get there.

The challenge was to avoid the airspaces of different countries that did not allow us to fly over them. Some would be easily avoidable, but the charts were also showing part of our route getting dangerously close to others. In addition, while some countries allowed us to fly over them, the diplomatic clearance given to us was only valid for a specific time. So we had to make it on time, or we'd have to coordinate for a new clearance. Between hard entry and exit times and other off-limits airspaces, we annotated the charts to make sure that we wouldn't be directed to a point that would be illegal for us to fly to. Flying around closed-off nations also meant detouring to get to our destination. The most direct path between two points would be a straight line, but for us, the path ended up looking like an M, and that required fuel to get there.

After calculating our fuel requirements, the plane was pre-flighted and everyone who wasn't flying prepared the back of the plane for the voyage ahead. For the flight crew up front, this would be the last time we started engines outside of a deployed environment for the next few months. This flight was to take us into the US Central Command area of responsibility (AOR), where the majority of America's fighting was taking place (Iraq & Afghanistan both lie in under that Command's geographic control area).

We started engines without incident and ran our checklist for our last non-combat flight we were facing.

E: Alright... Before take-off checks: trim tabs?
P: Set.
E: Flaps?
CP: 50 percent.
E: Flight controls?
CP: Feet and knees clear... Checked.
E: TOLD (Take-off and landing data)?
CP: Checked.
E: Crew briefing?
P: Ok guys, left seat take off.... If at any point before you see a reason to abort, call out 'reject' followed by prop, system, engine or load, and we'll keep it on the ground. Any time after that speed, we'll take it airborne and deal with the problem in the air. Engine shut-down procedures have already been briefed. Our two engine ceiling is currently subterranean, so if we lose an engine on take-off, go ahead and unlock the fuel dump safety pins. If it looks like we'll lose another engine, start dumping, don't wait for my order.

And the crewmembers took turns briefing the relevant information for the departure. The speeds required to achieve lift, the direction they needed to fly, the procedures in case they had to abort. Once that was done, the briefing was called complete.

E: ADI (Attitude Direction Indicator) mode and NAV SELECT panels?

P: I've got VOR/ILS 1 tuned the emergency return, pilot.

CP: Set, SCNS (skins – Self Contained Navigation System), Co.

E: Vertical Reference switches?

P: Set, pilot.

CP: Set, co.

E: Radar?

P: Set, weather, TCAS (Traffic Collision Avoidance System).

E: Radar altimeters?

P: 200 for the ILS, pilot.

N: 200, Nav.

E: GPWS (jip-wizz – Ground Proximity Warning System)?

CP: Set.

E: SCNS enhanced interrupted alignment?

N: 10 seconds to go… Complete.

E: Before Takeoff checks?

CP: Complete, Co.

N: Nav.

L: Load.

E: Engineer. Checklist complete.

Soon after, we taxied to the runway flanked by green fields of grass and took off, leaving for good the familiar Western World, heading east for Asia.

Contrary to the last time we took off, this time we launched in the morning when the Sun was up and would be lighting our world for most of our flight. This allowed us to view the world below us transform from a lush, diverse ecology of innumerable green trees, blue lakes and grass, to land where it was mainly great, open plains of dried, brownish grass. In correspondence, as our large warplane headed east and we left the popular West, the radio chatter and overall air traffic around us subsided, like the variety of trees and the frequency of towns and villages. The sentiment that we were leaving a populated world for a more isolated one was persistently reminded every time we crossed mountain ridgelines along our route. There were fewer and fewer population centers, major roads or aircraft sharing the skies with us.

To compensate for this feeling of loneliness, that the world around us was slowly disappearing, the crew started to converse more, knowing that we would all be sharing our lives for the next few months; from now on, we were each other's only source of company. More crew discussions occurred as well. Stories known as "so there we were..." (similar to what this entire story is) were recalled, different experiences eliciting laughs from the listening crewmembers.

It should also be noted, as noted above in transcripts from normal checklists being run, that the crew does not know rank when aboard the aircraft. Everyone assumes the responsibilities of their duties, without rank every being a factor. If the enlisted flight engineer doesn't like what the officer pilot is doing, he can still speak up and tell the pilot what to do to rectify the situation. It's meant to prevent what's called the 'halo effect' in the aviation community. A similar problem exists in the civilian and commercial aviation worlds as well, where an experienced pilot may make a mistake, but a brand new co-pilot doesn't want to anger his superior by pointing it out. The USAF has gone to great lengths to break down these rank barriers on the aircraft, successfully boosting communication between the crew-members.

For us, our training already got us ready and prepared to correct each other and call each other's attention in a formal setting. But that doesn't mean that the uncomfortable feeling of correcting an experienced member doesn't go away. Getting to know each other a little more on a more personal level, would help alleviate this uncomfortable feeling and prepare us for living together for long months.

Similar to the previous missions, we simply spent the long hours airborne deviating from some controller's headings in order to stay out of forbidden airspace, all the while talking amongst each other, slowly closing in on our first deployed destination.

Ten hours later, we landed in Belkastan. In the middle of the night, our aircraft was guided by an American airman to its parking spot, right next to our sister aircraft. Several hundred feet behind the marshaller, though, was the civilian air terminal of the capital city of this former Soviet republic. Parked for the night there were two Soviet-built Tupolev Tu-154 aircraft, rarely ever seen in Western airports. They were old and looked worn. Someone on the C-130 jokingly asked if these Soviet aircraft were perhaps permanently parked at the aging, communist-looking, blocky and massive concrete air terminal which had giant Cyrillic letters written on it.

Without any further discussions, everyone got their gear and hopped off the airplane. The maintainers were doing a final walk around with McKay, while everyone else piled into a white, American-made school bus. The Air Force airman driving it was another reminder that we were back on an American installation, with people here having already been deployed for several months. The Transition and Staging Port we landed at was just that: one of the United States' staging points before entering the warzone that was Afghanistan.

But the American base was still just a fenced off corner of this country's national airport. To get us there, the bus drove around walled off sections of the civilian world, specially built to hide the American presence here. We went through three checkpoints, two of them manned by Belkan soldiers. The third was manned by Air Force Security Forces.

It was only when we drove through the American checkpoint that we lost sight of the former Soviet republic. Before that, we were driven by old, decrepit and abandoned airplanes and buildings that used to be used by the Soviet Air Forces. The Stalinist Architecture buildings were still just as massive and monolithic as we had imagined. Trying to not to be stereotypical, many of us still ended up feeling like we were back in the 60s and 70s Cold War block that the United States opposed. There were many more Antonov and Tupolev aircraft that were decidedly abandoned, their cockpit windows shattered with some vegetation growing out of them. A Hammer and Sickle emblem placed above a nonexistent door of a building we came to imagine was a regional headquarters office building for the Soviet Air Forces only emphasized we had changed worlds.

But losing sight of the Soviet world as we passed the Hesco barriers topped with Constantine wire revealed another alien world. A tent city was erected in the vast fields that were now under American control. US forces had not taken over the old Soviet infrastructure, but instead were living out of tents and converted shipping-containers, one of them being labeled the Logistics Center and our first stop. There we got our keys for the tents we would be resting in for the night.

The Transition Center was an odd in-between installation. Military members were deployed in less than ideal conditions (sometimes even worse than some areas in Afghanistan which is in the warzone). But one could tell who was headed for Afghanistan and who wasn't leaving the Transition Center based on their uniforms alone. Officials working at the Transition Center wore the same uniforms as they did back in the United States, while Afghanistan had a mandatory uniform called the Multicams (it had many other names that don't need to be listed). The concept of uniformity was consequently thrown out; everyone could easily pick out who lived and who transited through the Center. And our crew was clearly transiting. We were placed in a section of the tent-city that was never to be developed into something more than just a temporary tent site.

Arriving at our tents, the flight crew was afforded accommodations that allowed for crew-rest, whereas others that were not in a crew-rest required job stayed in a different tent. The difference simply meant that aircrew were two to a plywood cell inside the large tent, while the rest slept in the same-sized large tent with fifty other people and no plywood walls. The tents had mattresses, but we still needed to walk to another tent manned by local nationals that worked on the American installation, where they handed out linens. We grabbed our linens, returned to our tent, placed them on the mattress, and then regrouped at the shuttle bus that was ferrying us around the installation.

Driving us to another prefabricated building, the shuttle dropped us off at a critical stop for our staging to Afghanistan. We stepped inside a large warehouse with rows and rows of fifteen foot tall shelves containing gear that was about to be issued to us. Two airmen gathered us in and started to brief us on what was about to happen. We were going to walk in a single file line through the aisles and grab a single item from each bin all along our path. We would grab everything, from gloves, glove inserts, flak vests, flak armor, helmets, jackets, boots... The list went on. To help us carry it all, we had our bags that we use to bring all of this stuff with us to Afghanistan. Different details were mentioned: don't drop the armor – if it falls vertically, it will break. This was not comforting to know since it was supposed to protect us from a bullet travelling 4,000 feet per second. The trick was that the armor was incredibly strong on the flat face, but if dropped on its sides, that's where the armor's weakness was (it would crack). In addition to our bulky protective gear and cold weather gear meant for the mountains of Afghanistan, we were also given our chemical warfare gear: gasmasks, suits, filters etc. Aircrew already had separate chemical gear to fly with, but this would be our ground equipment, if we weren't flying. The mask was easier to use – that was pretty much it. Chemical warfare for most in the military wasn't feared because of the terrible pain and horror the chemicals could inflict; no, chemical warfare was despised because it would require us to wear this gear. Each airman at the warehouse measured the gear on us as we all tried it on, and they adjusted the straps properly. The armor was inserted into our vests. Serial numbers were matched with our names and then we were permitted to take out our two huge, heavy bags outside to a loading pallet that would be forklifted onto the airplane. Another luxury of flying our own aircraft to the warzone was that we didn't have to lug this equipment personally – we placed it on a pallet and off it went.

And when it went, we were brought back to our tents, but instead of going to bed, we opted instead to head for another tent to get some food. The meal time was over, so we fed off of left-over fruit, muffins and Skittles. After drinking a bottle of water, we hoarded a few more for our sleeping accommodations before returning to our tents to prepare for bed.

After changing into our Physical Training (PT) uniforms (since wearing civilian clothing was no longer allowed), we all, at some point or another, went to the latrine tent, where toilets were placed atop a sceptic tank. This meant we had to climb up a few steps to reach the toilet before stepping back down to wash our hands. The plastic sinks used foot-pumps that fed water that was stored in a black bladder outside. This was nice because it meant that the water had all day to warm up; which meant the best time to take a shower was also in the evening. The morning would involve chillier showers, since the water would have cooled overnight. We brushed our teeth behind our sleeping tents using our bottles of water to rinse out the toothpaste; water coming from anything other than a sealed bottle being unsafe to drink. This would be true for the rest of the deployment. The last time we were able to drink water from a sink was this morning, at our hotel in Europe.

In addition to water from a tap, the last place we would be allowed to drink any sort of alcohol was here at the Transition Center. Afghanistan was a 'dry' country, which meant no alcohol was permitted anywhere in the country, and the US military obliges by that rule. There would be no option for us to leave the base to go get some alcohol elsewhere, although some crews would have the opportunity to leave Afghanistan on missions to other nations, and if they had time, to drink during their off-time. So for most of us, if we weren't getting ready for bed, we were walking to the Transition Center's makeshift bar that sold the last alcohol we would drink for potentially several months. Even at the Transition Center, All American military forces were limited to 3 drinks in a 24 hour period. Some of us gathered on the patio of the small shack with a drink. Some had foreign beers or wines, while others drank their last American beers or liquors for the next few months. An added bonus of the bar was that it was one of the few places that had Wi-Fi as well.

This was the first night we would all be hunting for reliable Internet connections. The time difference between the United States and our location was also growing wider. Whereas in Canada we could manage a few hours of difference with our loved ones work schedules and our own insane hours, the 10+ hours meant that now we had to work around not just work, but also sleep schedules and more. But at least we would be able to check our Facegrams, Twibooks and Snapsters. As we all gathered around small plastic tables, everyone checked their watches to make sure they were outside of their 12 hours (US Air Force regulations prohibit the consumption of alcohol within 12 hours of take-off), and then dove down into their phones. With the Wi-Fi connection established, most people stayed quiet, creating a scene that was eerily familiar to any American observer: friends out but connecting with the web instead of each other. The difference was that all of us would have hours of flying and then time in Afghanistan to get to connect. No one was in any rush to precipitate that, and the reality was that many still missed home dearly.

## Fourth Leg – Arrival at The Hub

Our sister ship had taken off a few hours ahead of us again and was well on its way to The Hub as we were getting ready to power up our own airplane.

Most of us had opted to skip breakfast – we were not in the mood to travel from tent to tent to go eat food that didn't seem very appetizing after taking a shower in similar-looking tents that were certainly a very different experience from our usual shower-styles of the Western world. Instead, we wanted to see if we could take off early. Grabbing our gear, the officers headed to the Command Post to grab their flight plan, while the enlisted went to the airplane to start preparing the aircraft for the flight.

Because we had no access to the tactical situation in Afghanistan, we were filed to fly under instrument flight rules to our destination, which meant that we would not be talking to any combat agencies, instead flying high above it all. We would be invisible to the war effort still, just like airliners that pass over the country everyday between Asia and Europe. With the information we needed to get to our final destination, the crew was ready to go relatively quickly. We had done this a few times by now.

The engines started, and we taxied onto the supremely long and wide runway. The Transition Center was where many of the Air Force's tanker fleet was based, and they enjoyed long and large runways. There were a few other bases outside of Afghanistan that hosted the flying-gas stations whose callsigns were EXXON and TEXACO, but none were stationed in the warzone. They would be too tempting of targets (airplanes filled with jet fuel) and didn't have a purpose being so close to the battlefield – all they carried was fuel for other airplanes. As long as one of them was available in flight to refuel a supporting aircraft, they could stay based in a safer area.

This meant that the flight was relatively uneventful again. As we flew across the other Asian nations toward our final country, the radios were quiet except for the occasional American flight flying in the opposite direction to us, bringing troops and equipment back to the Staging Port from Afghanistan. The silence as we flew over the ragged mountainous peaks reflected our moods. We were all quiet, coming to terms now that we were headed to Afghanistan. The silence around us also felt like we had reached the edge of the world. Prior to reaching this point, we always could have had the plane break down or something else happen that would delay our arrival in the warzone. Now we were committed. If anything happened, we were to push to Afghanistan. In the war-torn country would be proper maintenance facilities and facilities for us to live in for the next few months.

Of course, there was an entire unit waiting for us to go home. They were stuck in Afghanistan until we showed up to replace them. The military would not lose any capability simply because someone was late. If someone was late, the others stayed to cover the shift. It's terrible form to be late to a deployment, which is why the crews and planes were split. Most would arrive in Afghanistan on time. Those that stayed behind to fix the airplane would rush to catch up.

About ten minutes before we crossed into Afghanistan, we opened a checklist reserved for aircraft that were designed for military and combat operations.

N: Crew, combat entry checklist.

P: Acknowledged, pilot.

C: Co.

L: Load.

E: Engineer. Ingress and egress brief?

N: No egress. Otherwise, nothing has changed from the brief given on the ground earlier. Unless there are any questions, reviewed Nav.

P: Pilot.

C: Co.

E: Altimeters?

P: Set, 2-9-9-2, pilot.

N: 2-9-9-2, SCNS set, Nav.

E: Radar altimeters?

P: 200 for the ILS, pilot.

N: 300, in case of tactical low level, Nav.

E: GPWS (jip-wizz)?

CP: Set, tactical.

E: Unnecessary equipment?

P: Off, pilot.

CP: Co.

N: Nav.

E: IFF/TCAS is set... Defensive systems?

N: Ok the 24 is powered, program 4 is loaded, cocktails are set, and the ALE-47 is slaved to LAIRCM, which is...activated. Turrets are good. Defensive systems are set, on, Nav.

E: Defensive system safety pins?

L: Removed.

E: Survival equipment?
P: Donned, pilot.
CP: Co.
N: Nav.
L: Load.
E: Engineer. Exterior lights?
CP: Set, Co.
E: Engineer. Interior lights?
P: Set, pilot.
CP: Co.
N: Nav.
L: Load.
E: Engineer. Night vision goggles?
P: Ready, pilot.
CP: Co.
N: Nav.
L: Load.
E: Engineer. Lookouts?
P: Cleared to position.
L: Posted.
E: Combat Entry checks are complete.

The plane was configured and as ready as it could be. Our few passengers (the maintenance crew) in the back were told to take down their hammocks and get out of the litters installed for comfort. Since we were about to enter the combat zone, they had to be seated properly, just in case the aircraft had to be suddenly maneuvered away from a threat.

From the outside, the plane looked unchanged, which belied the two laser turrets on the aircraft that were activated by Carter and scanning for potential enemy missiles. There were none for now, though, and the aircraft continued to race forward, following the invisible air-route that few airplanes voluntarily took. Even with this checklist, we still felt alone in the skies, with no other signs of aircraft around us. From afar, our plane was a black dot passing over waves of desolate mountains.

Control: SHANCS 5-9-4, central Asian control. Your clearances are good. Switch to 123.4. Good day.

SHANCS 594: Switch to 123.4, SHANCS 594, good day.

The radio frequency switch was collocated with our transit into Afghan airspace. The area and nation's airspace were still very much guarded and overseen by the US military, which meant we had to be clear to enter the airspace like any other country. We typed 123.4 into our computers and after we pressed Enter, the entire dynamic of the flight changed.

Radio chatter exploded: innumerable aircraft talking to controllers, making the radios extremely busy. Flying over the border seemed like we had entered a loud, brawling pub from a quiet street. The activity in Afghanistan was a stark contrast to the world outside its borders. We checked in on the radio as soon as there was a free moment. We were greeted by American controllers, a sharp contrast in language compared to the broken English most of the neighboring Asian nations had been speaking. It was almost disorienting, and we had to respond immediately to free up the frequency for the next aircraft wanting attention. All other airplanes on the frequency were also speaking clear English (either British or American, or even maybe a slight French accent somewhere). But it felt like we had entered another busy American airspace, similar to Washington, D.C., Denver, or even Atlanta or New York.

We also activated a set of systems that had, up until now, remained dormant – but the new risks forced us to turn these on. We un-safed and activated our defensive systems just in case we were to become very unlucky as we approached landing. But at twenty something thousand feet, we were relatively safe, and below us was nothing but desolate mountain peaks. Our approach path right now had us flying in a less-accessible part of Afghanistan, which by extension was relatively safer. The real risk would exist when we would start to descend. But now at least, pods filled with burning hot flares and two laser turrets were armed to defeat most heat-seeking surface to air missiles. This was most of the flight over.

The controllers said hello and we were allowed to proceed on our merry way, their attention more focused in a different part of the country where the busiest airspace was being coordinated. Around us, there were just mountains everywhere and not many signs of civilization. It seemed like no one lived between Afghanistan and Europe and the view only emphasized that (ultimately inaccurate but still existent) perspective. An ocean of mountain peaks, not snowcapped yet, since it was still late summer, but nonetheless, barren and seemingly harsh to live in.

This was the first view of Afghanistan for many of us. The beautiful mountain crests and distant green valleys between these jagged peaks, banked by exotic looking villages, made us consider perhaps this place was, in fact, worth fighting for. The roads from up high didn't seem paved. And there certainly was no major highway or large buildings that suggested anything but a peaceful way of life. Ironic, given that we were here to help support a war.

The world seemed disturbingly peaceful, but the radios belied that impression. A-10 Warthogs were calling over the radio for permissions to reach a tanker through specific routings. Spy planes were coordinating for flight plans to their orbit points. The weather was apparently poor in sections of the country, which meant that many combat flights were not talking to the tactical controllers but the normal instrument-flight rule controllers to make sure they didn't run into each other mid-air.

About 20 minutes after puncturing through the northern border of Afghanistan, we were finally getting close to our destination. The weather was bad, and we couldn't descend ourselves to our destination; controllers had to guide us. They stepped us down, little by little, taking us south of the airport to then turn around and head back north toward the airfield. The mountains just to the north of the airfield were frustratingly high, and given the weather conditions, we didn't want to just dive down. Instead, we went south, and got in line to fly through the bottleneck airway that was mountain-less and the only route to The Hub. Everyone heading to the airfield was using the same one-way airway we were on.

After getting in line, the crew ran through its usual checklists. The descent checklist and before landing checklists were all accomplished. As we broke through the clouds, we got our first view of what would be our new home.

The view was bleak: city of containers and plywood shacks, occasionally interspersed with larger buildings and hangars, some left-over by the Soviets after their 1989 invasion and others recently built. The airfield was crawling with aircraft of all shapes and sizes. Helicopters from the Afghan National Army, built by the Soviet Union, were parked next to American Chinook helicopters taking off. Highly advanced C-17 cargo jets were parked next to tiny Sky Truck airplanes, and in the distance, the Special Operations compound had their single EC-130J, meant to ensure the United States had a psychological warfare capability. Descending lower and lower, we saw MRAP vehicles patrolling the fortified fence line (the wire) surrounding the base, guard-towers every hundred feet, and hardened entry control points.

The controlled chaos was normal and would continue for the next period of our lives. Radios were jammed to a point where it was difficult to understand tower sometimes. Dozens of aircraft were begging for attention as they all waited to land or taxi somewhere. Trucks loaded with fuel and others with bombs were waiting to cross portions of the flight line to tend to their airplanes.

A few hundred feet above the ground, the crew made sure the defensive systems were still off. They had been shut off when we had entered the weather to make sure lightning or other weather elements didn't actually activate the system, but a few hundred feet above the ground is the usual spot for us to safe them. We don't want to be spewing molten balls of fire onto the ground below us. By then, we crossed over the base's boundaries where the chances someone was waiting to shoot at us were lower; it was American controlled territory (more so than the overall grounds beyond the base wire that were generally under Afghan control with American help).

The first landing in Afghanistan was simply uneventful. Like most other very busy airports, we communicated and taxied to where we needed to go. We followed a tiny MC-12 airplane that was following a colossal C-5. All parking spots were surrounded by fifteen foot tall concrete barriers called T-walls (because they stood like upside down Ts). These walls were to shield other spots from explosions; if one aircraft was hit, the debris and explosion wouldn't affect the plane parked nearby. We would sometimes pass a set of walls and see there had been a fuel truck waiting for us to pass to then cut across to the other side of the taxiway.

Peeking over the walls were the different tails of the C-130s we would be flying for the next few months in combat. There was the familiar stripe our sister ship had, and then the tails from Japan and Little Rock were also visible. We found our spot and pulled in without incident. The engines were shut down, the doors were opened, and we all set foot on the dusty grounds that would be home from now on.

The major pace of operations was evident given how quickly we were ushered in to inprocess at the military facility. The old military mantra of hurry up and wait was reliably applied. Sitting in a small transit lounge, we waited to get our initial informational briefings – where the cafeteria was, where our sleeping quarters would be etc.

Martinez: Hi everyone. Welcome to The Hub. Sorry for the mess all around you, the building was hit by mortar fire a few days ago; we still need to paint over the repairs.

This was a disconcerting thought. The reality was still sinking in, but it also seemed so elusive. The same way an old war veteran tells you his story of a brush with death. You think this is just another "so there we were" story. The main concern wasn't how it was an actual possibility we would be hit by a rocket or mortar, but when we would be able to get to our rooms and finally go to sleep.

But the in-processing folks wouldn't let go. The first few briefings went by – reminder of the rules, the locations, and then the reminder again. A briefing that was oddly reminiscent of the duck-and-cover teachings of the Cold War, this one was simply: you hear the sirens, you dive to the ground. The likelihood of being hurt by an impacting round was virtually non-existent if we followed the simple rule: dive when you hear the sirens or the rounds. Then you wait. You keep lying down until the general PA tells you the attack is over or 5 minutes from the last impact – whichever comes first. And then you rush to the nearest bunker. And there you wait again for the all-clear. The end of an attack didn't mean it was all-clear. Teams still had to secure the bombed sites, the Quick Reaction Force (QRF) may not have stopped the attackers yet, and a full accountability had to be performed. So we would have to wait. If we were in our dorms, we were supposed to dive for the ground too. The commander of the base was very clear that this was one of his highest priorities.

We would later get more information regarding what to do if we were inside the aircraft once we broke off for our individual duty briefings. Until then, we were finally released to meet up with our First Sergeant. He had been on the first plane that had dealt with all these briefings earlier, so he had already gotten the ball rolling on securing keys to our rooms.

Like the disorganized gaggle we were, the last of us travelled in the night to the set of TEU containers that were merged to create a small building, each room for two people being a single container. While we had been briefed, terminal operations personnel had unloaded the airplane of our luggage, and trucks packed with our bags were pulling up to our container dorms. Before deploying, we had all been issued flashlights. Most of us had done the sensible thing and placed them in our bags... which we now had to find in the darkness. The few that had been odd and carried flashlights with them around the compound were our new best friends as we all tried to find our bags (which all looked the same, since the military likes uniformity). Within fifteen minutes, we had finally all individually grabbed our six bags and started to hunt for our rooms.

When we entered the rooms, we saw what we were given: the simple bedframe and mattress, along with a standard, high-school-like locker to stow our personal equipment we'd want to potentially lock up. Everyone but the commander was two to a room (which would be bumped up to three by the end of the deployment). The room was as wide as a bed was long, and as long as three beds put length to length behind each other. That meant that when two military members entered, it was crowded. And when they dragged in their 12 bags, it got tight. Crews got creative with how to stow their bags, shoving what they could under the beds, stowing the rest on top of the lockers or making furniture out of the remainder. Some people had foreseen this potential trouble for space and traveled with large plastic cases instead of suitcases, which turned out to be very practical. Over the course of the deployment, different pieces of wood from around the base would be collected to refurbish or build pieces of furniture such as a desk or shelf. Most of us had also brought folding chairs. These would be our chairs for the rest of our time there; until all you have is a bed or a floor, you don't truly realize how simple yet comfortable a chair can be.

At this time, a note will be added to emphasize how lucky we were to be living like we were. We had a solid wall and roof with a light that could be turned on using a switch. We had mattresses and a locker. It is fully acknowledged by all on the crew that the conditions faced by other members of the military may be more difficult. Living in tents, where the only private space is your cot and the space just under it. Our gratitude did not spare us from having to live in those other, more difficult conditions at the conclusion of the deployment. Regardless, we all acknowledged that having the luxuries we were afforded for most of the deployment were exactly that: luxuries.

Finally, we all ensured that our windows were plastered with cardboard and duct-taped shut. Not a single ray of light from the exterior world could be permitted to enter. The doorsills were also covered up. Crews were expected to sleep at any time of the day or night, and sunlight could be a nuisance to sleep (naturally). We would see sunlight only when we exited the building. Otherwise, the only source of light would be the pale white lightbulb provided by the US government inside of our metal box which we now called home.

Over the next few minutes, everyone unpacked their bedsheets and pillows, made their beds and quickly went to bed. When the roommates agreed to shut off their light, the room went dark.

And everyone stayed awake.

While it was dark and late in Afghanistan, and we were expected to show up early the next day at the Operations building for aircraft briefings and exams, it was still the middle of the day back in the United States. Which meant our bodies still thought it was time to be awake and busy. Some crewmembers swallowed a sleeping-aid pill, but others saved the pills for future use. We were only allowed a specific amount of pills per month in order to prevent addiction issues. We weren't flying the next day – just taking tests and doing discussions. Perhaps the pills would be more useful when facing a flight, when being alert and well rested was slightly more critical. Half the aircrew members were to regret that decision the next day though.

Irrespective of what happened the next day, we were finally in Afghanistan, falling asleep for the first time in our new home, and certifiably in a combat zone.

# First Flights

## Walk

This was it. Our crew's first, real combat mission; the culmination of all our training, designed to ensure that regardless of what happened, we were ready to handle almost anything. However, because of the combat environment specific to Afghanistan, details were different than what the crew had experienced in other operations across the world or in training. This meant we needed some quick introduction to those differences. For one day, we were assigned one extra crewmember from the previous rotation of deployers we were replacing. Then he would be gone, and we would be left alone. This first day was where we could ask questions and establish the basis for every day after – the same procedures and habits were fairly followed throughout the rest of this deployment.

Most of us had fortunately slept relatively well; jet lag was still a factor, but the times lined up well for us, where the first take-off was scheduled for 11:10 local time. The day prior we had taken tests to verify we were capable of flying in the combat zone and certifying we were still qualified to fly the airplane. Each crew upon arrival was also assigned a cell-phone that could connect to the local wireless network in order to keep us connected at all times to the Operations building, which was manned 24 hours a day, 7 days a week. We had received a few more welcome briefings and then proceeded to personalize our rooms. A few pictures were taped onto the walls and the beds were more appropriately made compared to the hap-hazard way we had simply collapsed onto our beds when we first arrived. We had unpacked our bags and soon enough, we found ourselves again going back to bed (exhausted from the lack of sleep the night before) to make sure we had enough rest for our first flight.

We woke up a half hour before alert time to get showered and dressed. At alert time, our crew's cell phone rang and the Assistant Director of Operations alerted us and activated us as a crew for the day. We had one hour to get to the Ops building for our briefings, so we all left to go get some breakfast at the Dragon Dining Facility (Dragon).

After a thirty minute breakfast, we left for the Aircrew Flight Equipment (AFE) cabin. The advantage of being based at The Hub was that everything was within walking distance for us. Our dorms, which were literally cargo containers welded together, were about a three minute walk from Dragon. Ops and AFE were both right on the flight line, also about a three minute walk, creating a nice little triangle where our life could be. It didn't take us long to get through the gates in the barbed wire fences to access the even more restrictive flight line, and go to the AFE cabin. Inside, we found neatly laid out on a table a set of Night Vision Goggles (NVGs), battery packs, laser-protective glasses, and rescue radios for each crew member. All of us grabbed our personal helmets and flight equipment from our individual lockers and then started to prepare and examine our NVGs at the table. We tested the night-vision goggles in a special dark room, tested the other systems, thanked the AFE team that prepared the gear and about ten minutes later, we were all done and ready to head to Ops.

We got to Ops early and met up with the extra crewmember we were replacing, called Guide. Carter went into the Intelligence and Tactics office to check out all of the mission secrets and computers for the mission. While he reviewed the flight plans, airfield diagrams and memory cards, Sheppard went to the Aviation Resource Management office to grab the radio secrets, which would allow the crew to communicate in flight. The loadmasters took a quick look at the requirements coming up and called the terminal at The Hub to tell them that they would be ready to load in a few minutes.

After all of these quick individual checkouts, the crew ultimately ended up a few minutes later inside the briefing room. The conference room table at the center of the room, and just large enough to fit the crew around, was placed in front of a large TV screen that displayed information from the connected computer. Strewn across the table were some papers that were just checked out by Carter and a large binder that held survival documents for the aircrew.

As we would do before every flight, we traded in our patches for a small pouch that held essential survival gear, such as a small map, compass, mirror, some fishing lines, and papers that could be used to communicate with the locals. The trade-in was designed in case we unfortunately found ourselves on the ground in enemy territory. If our luck didn't improve and we were captured by terrorists, at least we wouldn't be wearing patches that would identify which unit we were with and what our responsibilities were, delaying the intelligence gathering from the enemy. If we were fortunate enough to evade capture, we would still need some equipment to help us survive until rescue came. Hence the trade of patches for survival tools.

As we all settled into our seats at the table, an Intel member came into the room and gave us our intelligence brief. They went over all the information across all of our planned stops, briefing us the likelihood of enemy attacks or other info of the sort. After their brief, the Assistant Director of Operations would walk in to give us information about what is expected of us from the squadron. This usually was more administrative than anything; paperwork that may need to be submitted, a change in procedures or techniques, updates on our long term plans. Given that this was our first few days, the expectations had to be clearly presented though, so the usual 5 minutes were extended to 10. A crew had refused to complete a mission because the base had been attacked while they were sleeping, which had woken them up and broken their crew-rest. Leadership was clear that this was border-line unacceptable. The sacred rule of crew-rest was sacred (in case that wasn't clear), but it was also a privilege to be used responsibly – refusing to conduct combat missions because a base bombing had woken you up was not going to be useful since we lived in a warzone. Once the ADO left, the enlisted crewmembers left the briefing room to go pre-flight the aircraft. The engineer went to warm up and test every system, check almost every area of the airplane to make sure everything was in order. The loadmasters went to load the first batch of cargo and passengers after they completed their own checks of the cargo bay.

Meanwhile, the officers stayed behind in order to receive a tactical briefing. A tactics officer came in and presented information on the various tactics and techniques used to arrive at all the airfields we would be going to on this mission. There were also reminders from our training about how to react appropriately to different threats that may present themselves; for example, if we get shot at by a specific weapon, then our reaction should be this or that. After going over every landing zone (LZ) and airport, the tactics officer left for the rest of the aircrew to do a crew study.

Together, the three officers remaining discussed how they thought the day should go. Usually, there wasn't much left to brief, except for our first day, where we started assigning duties. Because of how busy radios were going to be and the lack of air traffic control of airplanes, it was decided that some radio calls would go to Sheppard, and others would be going to Carter. Hammond would simply listen to both at the same time and hopefully catch whatever the other two missed. Because the navigation systems on the airplane were difficult to use for the pilots, Carter would also have a laptop that gave him all the information regarding where they were and what portions of the country they were flying through.

When the crew study was over, the officers left the briefing room and headed out to the airplane with their four bags each. They carried secrets and survival gear, helmets, flight gear and more. The worse-off were the loadmasters since they also had to carry, in addition to their helmets and flight gear, cases of water and kits to help secure different cargo loads.

Ops was a small, narrow building tucked behind a large hangar, and just like most of the base, was surrounded by gravel paths. The flight line itself was fully paved and a change to walk on. After making one last stop at the Port-O-Potties at the edge of the tarmac, the officers finally walked out to the airplane to start their own pre-flight.

The flight line was always busy. As we walked on, the crew faced the eight airplanes assigned to our squadron at The Hub, although there were always two or three gone at a time. The aircraft were parked in two rows, head first, like your park your car in a parking lot. Between our two rows was a major taxiway that other airplanes used to get to their own parking further down. As a result, the crews always had to be alert to airplanes passing by, or a fuel truck rushing to refuel another plane, K-loaders or forklifts... It showed how active this war was, even if no one at home would ever notice.

The ramp and door of the airplane was almost always open when the officers arrived to the airplane, and the loadmasters already had the cargo loaded and ready to go. McKay was generally done as well, depending on how early officers showed up at the aircraft.

Getting into their seats, the officers started to turn on their own systems and perform their own pre-flight. Flight plans were programmed into the computers, radios were switched on, navigational aids were tested and defensive systems were verified. Compared to a traditional pre-flight in the US, these were relatively quick. A habit pattern quickly emerged after the first few flight, but unfortunately, during the first flight, the crew had a lot of questions. Thankfully, Guide was there specifically to answer our questions. Where in the US we would perhaps follow a specific procedure, we were told to skip it entirely in Afghanistan. Radio conversations were different as well. Before the engines were started, permission to proceed with the mission had to be acquired from headquarters (located in Qatar), and this was called receiving Positive Launch (PL – phonetically called Papa Lima). Carter would call over the radios for PL or use a text messaging service on the aircraft laptop to chat directly with Qatar.

SHANCS 55: Royal Love, good morning, SHANCS 55.
Command Post: SHANCS 55, Royal Love, go ahead.
SHANCS 55: SHANCS 55, request papa lima with sierra over delta.
Command Post: Copy, sierra over delta. Standby.

These code words were well known to the members speaking them. For example, sierra over delta could mean 4 over 2, where sierra was a code word for 4 and delta for 2. This message, relayed to headquarters in Qatar, told our controlling agency that we had four passengers and 2 pallets. If that was what headquarters expected us to carry, they would return a coded message saying we could proceed.

A few minutes later we received approval, the engines were started and the mission began. There were of course times when the crew started engines without PL, through omission or because we needed to.

When engines were started, the plane had to be backed out of its parking spot. Unlike in a car, the pilots could not simply look into their rearview mirror or through the rear-window to see where they were going. So with the ramp and door in the back still open, the loadmasters told the pilots how to back out.

P: Okay loads, as usual. We'll have our feet on the floor to make sure we don't hit the brakes. You keep talking, we keep backing. If we don't hear from you anymore, we'll use throttles to stop our movement and figure it out from there.

L: Roger that.

P: OK, we're ready.

L: Roger, come on back... keep coming back... keep coming back... Ok you can start your turn towards 4...turn towards 4...keep your turn...

The airplane started to turn as it continued to back out of the concrete T-wall surrounded spot.

L: Good turn... keep coming on back... less turn... keep coming on back... 5... 4... 3... 2... 1... Stop.

The plane came to a stop.

P: We got it up front.

L: Copy. Ramp and door coming closed.

The plane started to move forward, down the taxiway. This ritual was performed every time we left The Hub, and the loadmasters would tell the pilots to keep moving back, to turn towards one engine or another, and if they stopped talking, the plane stopped moving. Brakes weren't used because if the brakes were suddenly hit while moving backwards, it could throw the loadmasters out of the airplane. Throughout our time there, there had not been a single incident from this procedure. While it's expected, it's done so often that it's also expected that something would happen at least once through sheer repetitive actions, just like car accidents happen (if not with our crew, then with someone else's).

Ground: SHANCS 55, taxi to zero nine zulu via delta, foxtrot.

SHANCS 55: Delta foxtrot to zero nine zulu, SHANCS 55.

This is where Guide again showed how useful he was. To find your routing at an unfamiliar airport that was bustling with activity was a challenge on its own, but now there were tall, protective concrete T-walls everywhere that limited the view of our surroundings. Fortunately for Guide though, he was so used to the procedures that he simply walked us through it to where we could keep up and we weren't slowing down the operations at the airfield.

Our plane taxied toward the end of the taxiway that acted as a runway. Just parallel to this taxi-runway was the real runway, populated by large earthmoving and paving equipment and a large X made out of lights, always on to visually tell approaching aircraft that this specific runway was closed.

But as we continued to navigate our large airplane through the maze of T-wall barriers hiding other planes and aircraft at every corner, Guide reminded us that there was still more to do.

"You should run the Combat Entry checklist now. Unlike in the US, you're not flying into the combat zone – you're in it already. Better to have all the systems up and running the moment you take off. After all, you never know if someone's going to try and take a shot at us right as we take-off."

With that rousing speech of encouragement, the crew ran the combat entry checklist and prepped the airplane for the combat airspace.

When we got to the runway and switched to tower's radio frequency, we were welcomed with an unusual message. The Hub was under attack and everything was shut down – all aircraft on the ground had to halt and all arrivals and take-offs were halted. We, along with a few other planes, were stuck as rockets were being fired at the airport.

Sitting in the airplane, we didn't have much to do except wait and hope none of those rockets were going to hit us. Every few minutes, someone would come on the radio and remind everyone that the runway was closed and the base was under attack. We couldn't hear outside, such as explosions, under the buzzing of our own engines, and the only things we could see was outside our front windows, which all in all, was a fairly limited field of view. Even under this attack, our crew was surprisingly uncaring and unaffected by the fact that explosive rockets were being launched at our base. These attacks would be common place for our entire time there, and with a few exceptions, the crewmembers generally did not display fear or concern.

Twenty minutes later, the base was reopened, and it soon became our turn to line up onto the runway and be cleared for take-off.

At the end of the runway, with all the checklists complete, the throttles were pushed up and the plane started to roll for the crew's first take-off in Afghanistan this deployment.

After reaching one hundred and nine knots, the plane aimed its nose high and the aircraft climbed into the skies. Carter and Guide both immediately stood up and looked outside the front cockpit windows as soon as the plane had taken-off, not to observe the view, although it was beautiful, but to scan for threats. Historical analysis showed aircraft were most vulnerable when taking off and landing because they were flying slower and low enough to the ground that even a man with a gun saw an opportunity. But there were no issues on this take-off, and as they climbed away, the view of the world outside the fences of The Hub came into view.

Surrounded by tall mountains, the area was a lush green, with various gated houses located all around. Roads crisscrossed the green fields, small lakes were spotted the area, and the lifeless mountains in the distance were a reminder of the harsh life that faced the people below. There were no large malls, supermarkets or other structures that had large parking lots. That's what was different from what we usually see flying over the US: there were no oceans of concrete that served as an enormous parking lot for a Target, Wal-Mart, Home Depot, etc. It simply looked like an endless sea of clumps of houses lost within the fields.

Reality set back in when we were told to contact a different agency on the radios. As we switched frequencies, it was barely possible to hear anybody over these radios. The jamming designed to prevent explosives (IEDs) from detonating were also interfering with our radios. We could barely hear the controllers and they could barely hear us.

<u>SHANCS 55:</u> Departure, SHANCS 55 passing six for seven thousand, runway heading.

<u>E:</u> After take-off checklist.

<u>CP:</u> Complete Co.

<u>N:</u> Nav.

<u>L:</u> Load.

<u>Departure:</u> SHANCS 55, say on course.

<u>E:</u> Eng, checklist complete.

N: 1-2-0.

SHANCS 55: 1-2-0 for SHANCS 55.

Departure: SHANCS 55 turn on course, climb and maintain one three thousand.

SHANCS 55: SHANCS 55, on course and out of seven for one three thousand.

The crew had to balance multiple things at one time: checklists to get the plane set up, radio frequencies, talking to anyone listening on those radios, etc.

We turned the plane to the heading we needed to fly and continued to climb quickly. The Hub was *conveniently* placed in a geological bowl, which meant we had to turn away from mountains that were around us, climbing as quickly as we could. We would have to hop over those mountains soon enough.

All of our combined ears could make out what the controllers were telling us: there was a drone nearby and we had to avoid an airspace bubble. But we were then told to turn away from where we were trying to go, and radios again were breaking up, which meant we didn't know if we were allowed to turn back on course. To add to the problem, we still had to fly the plane and run our checklists. Using our radar and other computer systems, we stayed clear of other aircraft and the mountains as we continued to climb up to thirteen thousand feet.

Throughout our own challenges, other planes were on the radios, trying to get their own set of instructions. The jamming made it difficult to catch our call sign and make sure the call was for us.

Departure: SHANCS 32 turn left to two nine zero, traffic is a Beechcraft eleven o'clock at ten thousand.

That call wasn't for our aircraft – it was for someone that had the same name but different numbers. Normally, it's easy for us to make the difference, but with the heavy jamming, we did struggle.

SHANCS 32: OK, turning to two nine zero and we have that traffic in sight.

P: Descent checks.

Departure: SHANCS 32 maintain visual separation from that traffic, maintain heading to intercept final, contact tower 123.4

E: TOLD.

SHANCS 32: 123.4 for tower and we'll stay clear. Thanks.

CP: Checked

SHANCS 01: Approach, SHANCS 01, flight level two four oh. Request.

Departure: SHANCS 55 turn to a one eight zero.

SHANCS 55: One eight zero for SHANCS 55.

Departure: SHANCS 01 ident, state request.

SHANCS 01: SHANCS 01 with the flash, request vectors ILS to The Hub.

Departure: SHANCS 01, radar contact, 10 miles south of Gobble, 24,000, and you can expect that.

Sometimes we would respond to something that wasn't for us and further annoyed the controllers. With the airspace extremely busy, we had Guide helping us confirm when it was our call sign, SHANCS 55, being called by controllers.

Finally, we were turned back on course, and then we were switched to another controller – probably much to the previous controller's happiness.

Departure: SHANCS [garbled] course, contact approach [garbled].

SHANCS 01: Confirm that was for zero one?

Departure: Negative. SHANCS 55, turn on course, contact approach one twenty three point four.

SHANCS 55: On course and twenty-three-four for SHANCS 55.

The switch happened as we were flying over a mountain ridgeline, which meant that in addition to the radio jamming, mountain peaks were blocking the signal to the new controller, so establishing communications was even more challenging. We kept trying to establish contact, but no success. We could hear the airplanes around us talking, but we couldn't be heard, and we were now getting closer to the destination.

We finally established contact and were given different directions to get closer to the airport and begin our approach while dodging other aircraft that were flying around to their own destinations. To add more radios to our communications nightmare, we were also contacting the airport to tell them we were arriving with different types of cargo and passengers. As we were quickly learning about this country, they were also extremely difficult to hear and talk to.

Complicating this set of communications was the fact that they were not native English speakers, but European, so the accent over the radios made it difficult. Their vocabulary was fine, but the jamming and different intonations made it hard to understand what was expected of us. We finally simply said: "we'll call you back when we land," and focused instead on the flying portion, which was getting closer to landing.

Approach: SHANCS 55, cleared visual to runway 04, switch to tower 132.5, SALIENT 02, immediate left turn to 1-2-0.

SHANCS 55: Cleared visual to tower, SHANCS 55.

SALIENT 02: 1-2-0, no delay, SALIENT 02.

Switching our radios to talk to tower, we flew over a specific area that jammed so bad that Hammond, flying the airplane, punched off his radios and asked us to relay anything important to him. As we continued our approach, other planes were still talking on the radios, which made for a lot of background noise. The crew was reading different numbers to the pilot flying to feed him as much information as they could to keep his eyes on the long runway he was aiming for. Airspeeds and altitudes allowed him to adjust his approach, but the radios were so deafening that those reading the information were unwittingly shouting it to Hammond.

E: Speeds are 129, 109, 99.

N: 500 feet, one mile.

P: Did we get clearance?

CP: We are cleared to land.

When "10 feet" was called, Hammond pulled the throttles back and aimed the nose high to flare. The plane settled effortlessly onto the runway. We had finally landed!

We taxied off the runway at Gobble and got in touch with the air terminal that was waiting for our passengers and cargo. The radios were blissfully normal and we received our instructions to go and park at the proper ramp. A power cart and marshaller would be waiting for us.

After a few minutes of taxiing and running our checklist, we shut down our engines. Thankfully.

The flight had taken twenty minutes (it would later take only 13). But the absolute chaos and inability to talk had been shocking and exhausting. The crew still had seven and a half hours to go. Guide, who was showing the crew the ropes, also made it clear that this was normal and we would get used to it. This was our first flight: disorder, confusion and general commotion. While our initial feelings upon arriving in Afghanistan were those bordering naiveté, this was confirmed after this first leg. Some of us were starting to stress out about how potentially frustrating this deployment was going to be.

As our cargo doors opened, there were representatives waiting to grab the cargo and passengers. We had about an hour on the ground, but because we were late, we were trying to get back on time and rushed everything we could.

After reloading the aircraft with more cargo and people, the engines were restarted and we took off from Gobble for our next destination, Kandyland.

And the radio jamming welcomed us airborne as we climbed away. As we moved further south from the large cities, the jamming weakened and we were able to establish contact with Tetra.

Tetra controlled all the airspace in Afghanistan except over major cities. They knew where the tankers were orbiting, the spy planes were flying, where fighters were bombing etc...

But the radios could sometimes be monitored by some clever enemy fighters, which meant we spoke in a common, coded map-language. The combat airspace was divided into what are called killbox-keypads, dozens of equally sized squares divided the map. Each square had a number and letter associated with it; the first two digits and letters were the killbox. The last digit was the keypad, which were smaller, equally-sized rectangular divisions of the killbox, and they were numbered in the same order as a phone keypad – hence, keypad. So a big box was called a killbox, and that contained 9 smaller boxes, called keypads.

In addition to sowing confusion among unwanted listeners in case our radios were not encrypted, it allowed everyone to follow everyone else's activities clearly and controllers to easily block off airspace whenever needed. Instead of trying pass along four sets of coordinates with six or seven digits each to every aircraft that checked in, the controller could simply say one of the boxes was off-limits. Or a whole killbox, which closed 9 boxes.

Because of this, we always checked in with Tetra to ensure we weren't going to be flying in what was called 'hot airspace' – airspace where fighting was actively taking place. To the uninitiated, here's what it sounded like.

SHANCS 55: Tetra, SHANCS 55.

Tetra: [Static] 55, go ahead. [Static]

SHANCS 55: SHANCS 55, four three hotel delta four, destination one niner sierra delta six, flight level two four oh.

[Static]

Tetra: SHANCS 55, radar contact. [Static] …rra delta six via [static] seven, maintain two three [static] until then.

SHANCS 55: Tetra, you came in broken-unreadable for SHANCS 55, say again please.

Tetra: SHANCS 55, you are cleared to destination via seven two echo charlie seven, maintain two three oh until then. [static] Glasseye at [static] twelve thirteen line.

SHANCS 55: OK SHANCS 55 copies: two three oh until seven two echo charlie seven, Glasseye at the twelve thirteen.

And with that, we had to find out where in the world (literally) they wanted us to go (box number seven-two echo charlie seven), and which points they wanted us to switch frequencies (to talk to Glasseye). Like all other airspaces, we had to fight for some attention with other airplanes, but an hour later, after flying through the required areas, we landed again.

We were getting hungry, so while the plane was going to be unloaded and refueled over the next hour or so, two of the crew went with Guide to be shown around the airport and where to get food. There was a large Boardwalk square around which were a few restaurants, including a pizzeria and TGI Fridays, but everything was closed at the time we were there. As we headed back to the aircraft, the base began being mortared and we immediately took cover in a bunker.

Taking cover in the reinforced concrete bunkers, the crewmembers looked at their watch and saw their timeline slip away from them. There were cover bunkers located all around the major bases in case this eventuality did occur (which was often). When hearing the initial indications of an attack, we would hit the ground and lay on our stomachs for a few moments. After that, we'd make a beeline for the nearest bunker or reinforced building we could find. From there, we would wait out the attack as the forces on standby for this reacted.

But our day continued on. The chaos ensuing from attacks meant we were delayed again because fuel trucks did not move while we were being bombarded. When we finally were loaded up and refueled, we raced through checklists and proceeded to get airborne as quickly as possible again.

Airspace control was difficult to communicate with and the variety of airfields we went into meant we had to perform different techniques every time. Best of all, the next day, we would be alone with no one who's had any experience with this. But with each hop, we were becoming more and more proficient, and by the end of the day, Guide was simply sitting on the bunk, quiet, thankful he was finally done with his own time in the war zone.

Seven and half hours after our first take-off, we landed back at The Hub. Ten hours after we first woke up. Guide simply said good job and bade us farewell as he headed back into his tent, while we went to Ops to drop off our classified material and other gear and paperwork.

## Run

The next mission was going to be just us: the basic crew. The six of us had managed to leave our home base in the United States by ourselves; how tough could this be? Guide had shown us the ropes, right? This was going to be a pretty easy day. We were first flying to Sabarat, a small Landing Zone (LZ) south of The Hub, which would take about 40 minutes or so of flying. We would be performing an Engine Running Offload/Onload (ERO) over there.

That meant we wouldn't even shut down our engines at Sabarat, and instead, everyone and everything would be transferred from an airplane with its engines running.

Looking back on the day though, there were moments of nervousness that were of course hidden and never discussed amongst the crew. But we were all novices to the warzone. Our confidence in the United States or elsewhere in the world was born out of the thought that we usually had a good crew that had a mix of experience and learning. Not to mention that we were now flying in a country that was considered a combat zone *everywhere*. It felt like the first time you had ever driven a car by yourself. You had driven with your parents for some time and thought you knew what you were doing... But the reckless driving didn't come until a few hours after being by yourself. This small injection of insecurity was exactly what the crew needed. We would all consequently be hyper-alert, double check all our work and destinations, check each other and probably be the safest flyers we had been in a long time.

We started the day like we did the first time. We weren't ready to amend a procedure that had worked, so we woke up 30 minutes prior to getting the alert call, and then went to breakfast once the call came. After eating, we proceeded to AFE to get our flight gear and then to Ops to receive our briefings and take a look at our day. We were going to be supporting an Army mission. So the Army virtually owned us for the day; if they needed us to go somewhere, we would do our best to make it happen. Headquarters in Qatar would pretty much be hands-off unless they needed to coordinate something for us.

After the briefings, where we discovered new information about the LZs we were destined to land at, the enlisted members of the crew went to the airplane and the officers stayed behind and talked about flying techniques. The officers had to consider different tactics to minimize the threat from potential enemy fighters that wanted to shoot down an American aircraft, so they quickly discussed some of these tactics from our repertoire of skills we all had trained for in the United States. There was no reason to reinvent the techniques we had trained for, but instead just select a play and run with it.

The enlisted, instead of worrying about those things, went and loaded the aircraft and performed the preflight. The maintainers talked to the Flight Engineer about the status of the airplane. They had just replaced a temperature gauge for the number 3 engine and the hydraulic fluids were topped off. No issues, it was Alpha 1 – ready for flight.

Contrary to the day before, after having started engines and gotten to the end of the runway, the plane took off on time, and the mission was off to a good start. Radios were still just as catastrophic to understand, but Tetra was easy to deal with and allowed us to fly directly to Sabarat.

Sabarat was a small LZ that was scheduled to be closed within a few months, part of the US war effort winding down in the region. As a result, we were going to be carrying a lot of cargo and people out of this small strip. Some of the challenges facing the crew were that it was an inclined landing strip, so it would affect our take-off and landing performances. This resulted in the crew having to perform what was called an assault landing (and takeoff later). Procedures were different, and the crew discussed it several times on this flight since it was our first assault on this deployment. Guide had been with us on a mission that took our crew to big, regular runways, which meant he hadn't been able to present us tricks regarding these small, 3,000 or so feet fields. This was our first time flying into a field in Afghanistan that had very limited space, and stopping the airplane that weighed hundreds of thousands of pounds was going to be a challenge. We all needed to be on the same page. The area was very limited in space, so you had to nail the landing or go around and try again. Even details regarding parking had to be considered: the limited space meant you were asked to park a certain way to allow one additional aircraft to land; otherwise, when you landed, you shut down that little airstrip – the lifeline for the FOB that was surrounded by roads that could be laden with Improvised Explosive Devices (IEDs).

Doing some quick math in our heads, the crew decided to start their descent at a specific distance so that we never stopped descending from twenty thousand feet until we were feet above the runway. Pulling the throttles back and dipping the left wing, the pilot maneuvered the plane into a descent that increased our speed exponentially, and that initial drop even created a temporary zero-G effect, which most crewmembers enjoyed. We were in a controlled fall toward the Earth.

Checklists were starting to run quickly, and we were setting ourselves up for a single turn to end up lined up with the runway. The rate of our turn had to be finessed to ensure we lost the appropriate amount of airspeed at the right time, before we could lower the gear and extend the flaps.

Thousands of feet lower, without once having touched the throttles, the landing gear was lowered, and the plane was aiming at a tiny sliver of a runway, where a giant black spot at the end of the runway showed where all the planes before us had landed and left their rubber marks. The rest of the runway was clean, making it clear that if the aircraft did not land on that black spot, we would not stop in time. There were no black marks on the other end of the runway, which meant that this runway was unidirectional. This was a one way, one shot touchdown landing. If you didn't land 'in the zone', and you still tried to land, you would crash headfirst into a series of guard towers and get tangled in the cable holding down an aerostat floating above the runway. If you didn't hit the zone and decided to try again, you still had to climb and turn around to avoid those obstacles.

Hammond nailed the first assault landing. But wheels on the right spot on the ground didn't mean it was over. The plane roared down the runway, but the engines were soon placed into reverse thrust, and the brakes were applied, slowing down the cargo airplane to a slow walking speed comfortably at the end of the runway where we turned off of it and stopped. The sensation of time was odd – it took as much time for us to travel down ¾ of the runway as it did the last quarter, given the speeds we landed at.

The landing complete, the crew immediately started to prepare for the next set of objectives: expediently unloading and loading cargo.

E: Ramp and door?
P: Cleared to open.
E: Downspeeding two and three.
P: Understood.

Leaving the propeller engines running and still turning at 1,020 rotations per minute (or 17 times in a single second), the offloading and on-loading was beginning only a few feet from the runway we had just landed on. Keeping the engines on meant that the turn-around time was going to be quick: McKay needed to calculate takeoff information quickly as soon as the Loadmasters gave him the weights of the cargo being on-loaded, PL had to be coordinated for and approved, and the Loadmasters, of course, had the hardest job of all.

While one of the Loadmasters guided a forklift to the back, close to the ramp, the other would speak to an Air Terminal official about the cargo that was about to be on-loaded. Over the loud engines, the Load would quickly reconnect his headset, tell McKay what the weights were going to be, and just as quickly disconnect again to help the other Loadmaster with the offload.

Together, they grabbed pallets and signaled Carter, standing in the front of the cargo compartment, near the stairs to the flight deck, next to the aircraft locks system. It was there that the pallet control system was located, and the loadmasters signaled which pallets were to be released. Once released, the two Loadmasters would push the pallets off. They often found themselves struggling against the weight of the container-sized cargo because the plane had been parked inconsiderately on an incline that worked against them. But these airmen remained professional, and instead of taking the time to complain, heaved against the heavy cargo and pushed the cargo off the airplane.

Even in the heat of the engine exhausts and the sunny desert, they continued to push. When passengers came aboard, the loadmasters quickly got straps and bundled the luggage together to secure it. Of course, vehicles were even more annoying. The loadmasters slid beneath the vehicles in places no human is supposed to fit, with heavy chains to wrap the vehicles and secure them to the aircraft cargo floor. This was, of course, after they had to guide another person driving the vehicle backwards into the cargo bay.

All of this while they knew up front, the rest of the crew was sitting comfortably in the air conditioned cockpit.

After swiftly completing all of these tasks, the loadmasters returned to the radios and declare to the crew, with still some breath to spare:

L: Ramp and door: closed and locked, cleared to taxi.

On cue, radio calls were made for take-off and the plane inched forward to line up with the runway. For this first assault take-off, the plane sat at the end of the strip for a few extra minutes as the crew discussed in detail the procedures and considered what to do if something went wrong. This plane is heavy, we are go oriented, we will be timing for fifteen seconds to make sure we're gaining speed as needed. If we're not, we'll abort the take-off. If we lose an engine on take-off... And the discussion continued for a few minutes before it was time.

We were still learning about each other's techniques and comfort zones, but by the end of the deployment, these briefings would see the crew simply say 'standard, any questions?'

The throttles were pushed up while holding onto the breaks. The engines continued to spool up to their max power, blowing dust behind the four massive fans that were the propellers.

Then, the breaks were released. Like an accordion wave, the passengers in the cargo bay were jolted into their neighbor as the plane lurched forward and accelerated to take-off.

Seconds later, the speed was verified as being good and we were committed.

CP: Go!

After that critical call, the plane's nose aimed up and the cargo aircraft jumped into the air, sacrificing speed for altitude. Fifty feet above the ground and above any obstacles, the aircraft was maneuvered to a more stable pitch where it could still gain speed and climb, avoiding any risk of stalling the aircraft. The aircrew was recovering airspeed to use in case we were shot at. This would allow us to climb quickly again, or to turn suddenly without crashing due to a loss of airspeed.

Again, no such shots were made. The plane rose to its altitude, the crew struggling to reach Tetra as usual, but soon en-route back to The Hub, with the cargo that had to be moved.

Twenty minutes later, through the chaotic communications, the plane landed back at The Hub, finishing one the shuttle flights of the day.

After shutting down engines, offloading and on-loading cargo, refueling and getting some snacks, we were back in Ops. While in Ops, though, we discovered that we were returning to Sabarat. The original second stop, Sylvereal, was inaccessible due to rain damaging the dirt runway we were supposed to land on. Without delay, the crew went back to the plane, restarted the engines and headed back to Sabarat. We had to stay on timeline for the next few stops we had planned, and it wasn't difficult for the trained crew to react to these changes.

We took off, flew, descended, landed and did the offload/on-load movements again, gaining more confidence in our capabilities for this deployment with every minute.

Ten minutes after taking off from Sabarat, the crew was able to reach terminal controllers close to The Hub. The news was surprisingly not what we expected. We requested to land at The Hub, but were instead told that the runway at The Hub was shut down due to an aircraft emergency on the runway. Fortunately, the crew knew that Gobble was right next door, literally six minutes away, so we instead requested to divert to Gobble until The Hub would reopen.

But Gobble was closed, too. That airport was under attack from insurgents mortaring the airport. That news suddenly limited the crew enormously. The fuel load had been very limited to go to Sabarat. Too much fuel would have meant we would be too heavy, and would not be able to maximize our cargo load leaving such a small airstrip.

The closest fuel stop was Klein Deutschland, and that was an hour away. The little landing zone that was Sabarat did not have fuel, and it was not a good idea to just land there and wait with the fuel we had. Would the runways at The Hub and Gobble be closed for an hour? The controllers didn't know – no one knew. If we waited and the runways didn't reopen, we would be in trouble, and wouldn't be able to make it to Klein Deutschland.

The decision was made to divert to Klein Deutschland without delay, and even then, the fuel reserves were dangerously close to minimums. There was nothing between The Hub and Klein Deutschland except for mountain peaks. Nothing else – it was a mountain chain that stretched for an hours' worth of flying, which on a normal day, offered great respite (more in Chapter Peaceful Nights).

But as we started to navigate our way to the northerly destination, some on the crew got into the books to find out how to maximize fuel, just in case the winds would be tough to deal with or, with our luck, an engine malfunctioned or had to be shut down. In addition to checking our most fuel-efficient airspeeds and altitudes, the crew also spoke to the airspace controllers to tell them what our routing would be due to our newfound difficult situation. The initial symphony of procedures to come up with an alternative plan and check that we were going to make it were rather tense, each person looking at their systems to see how to they could maximize the fuel we already had (except for the loadmasters, who would have nothing to do but throw things out the back).

Fortunately, there weren't any problems. The aircraft performed well, even after bashing it into a small LZ with low fuel, and the airspace was not active between our positon and Klein Deutschland. After reaching the halfway point over the ocean of mountains to our destination, the crew saw they would land with a little extra fuel, which was a much better feeling than seeing we would land with no fuel.

We approached the airport and landed after the Sun had set. After taxiing off the runway and parking in the cover of darkness, we could tell we weren't entirely expected by airport authorities. Like most of the people across the country, most were likely busy getting ready for a quiet night, not preparing to fuel an American warplane that decided to show up unexpectedly.

But the base locals were still very accommodating, and after waiting for a few minutes, they drove a fuel truck for us to refuel from.

McKay, clearly tired from the long day that it had been, told the fuel truck operator to keep pumping fuel into the airplane. But someone else forced him to look at the gauges again. McKay then noticed that they had taken thousands of pounds of fuel more than they needed. Thankfully, it could all still be used and wouldn't be a waste. However, if he hadn't been stopped, the crew agrees that we could have probably flown to Alaska and back. Or more realistically, caused damage to the tanks.

The laughs resulting from this tiny oversight allowed to crew to be in a much more jovial mood on our way back to The Hub compared to when we had to divert and worry about running out of gas. We contacted Ops through our satellite linkup, and they notified us that the airport was reopening, and we would be able to return. And even if it were to be closed again, McKay guaranteed that we had enough fuel to orbit over the airfield until it reopened.

An hour later, we arrived over The Hub's airspace, and the controllers thankfully confirmed that the airport was open and we could land. But the challenges weren't over. Weather decided to join us over The Hub, which meant there were clouds everywhere. Our descent was much slower than what it was expected to be. The Hub sat in a geological bowl, surrounded by mountain ranges that reached up thousands of feet above the runway, which meant we had to use our radar and other Navigational Aids (NAVAIDs) to guide our way down through the narrow space we had. Being in the clouds after a long day was slightly disorienting.

The crew had to communicate clearly to each other. Looking at the radar screens, altimeters and charts, the crew was able to spiral down between mountain peaks, adjusting for winds that could have blown us into a mountain side. We could tell we were getting lower as the dark shadows behind peaks on the radar grew (the mountains were blocking the radar emissions from seeing what was behind).

When the runway came into sight, we realized we were too close for how high we were. We needed to get down quickly, so a rapid descent was initiated. Alarms started to sound inside the cockpit, and a voice blared "Sink Rate". We were descending very fast toward the ground that was quickly getting closer and closer – the computer systems aboard were not comfortable with the situation and continued to initiate alarms that we noted and then silenced.

But Sheppard landed it, even if it was slightly disorienting to him. Hammond observed from his seat and after the landing gave him a pat on the back with a welcomed "good job." The challenging descent in the dead of night and thick clouds was over.

After parking in our spot, we shut down the engines for the last time of the day. Our delay going to Klein Deutschland and back meant we weren't able to finish the original mission due to duty day limitations set by regulations. So we grabbed our gear from the airplane and walked back to Ops to finish the paperwork.

Our first flights were over. We learned to walk with Guide and had to run on our first solo mission. After having endured fuel diverts, base bombings, assaults, changes in mission plans and destinations, and weather challenges, our learning to drink water from a fire-hose was over. We certainly felt ready.

## Stop

After our first few days of flying, the crew's confidence was soaring. We were managing all of our operations on time and safely. We were learning to read the tone of each other's voices, predicting each other's moves and becoming impossibly efficient at our work. The crew was meshing well, too, on a personal level. Everyone was able to remain friendly and those teams that worked closely with each other for various phases of flight were also getting along (for example, loadmasters work together during the loading/unloading of the warplane).

Like at home, we still were able to multitask even with the terrible communications and challenging landings or take-offs to plan and execute. We talked to each other calmly over the intercom while checking over numbers or charts, never needing to make eye contact. We would be shouting at each other like raging gorillas if we didn't talk over headsets and the intercom, given how loud the plane was. If we were busy talking on the radio or simply didn't want to talk to someone on the plane, we could flip a switch that would light up the appropriate system and another crew-member would understand that the task had been accomplished. Or we could just give a thumbs up – that also worked pretty well.

We were considered a 'solid' crew and a capable asset for the squadron, like most crews.

But not all.

Another crew was being scrutinized by higher levels of leadership after a few weeks of flying. There were concerns about the safety and effectiveness of the crew. The squadron commander and his deputy were not in a position to offer that crew maybe 'easier' missions until they gained a good footing on operations. Unlike in the US where a different tempo for operations existed, headquarters in Qatar was told by our squadron they had X-number of good crews and HQ had to plan missions based on demands and requirements, often time-based requests. There were never enough planes for all the missions required which meant what they had was to be maximized as much as possible. Headquarters couldn't start planning operations on specific crews. The squadron also couldn't allow crews to cherry pick the sorties they wanted due to crew-rest considerations, efficiency and convenience concerns. Headquarters had a plan that fit into the larger operational picture of the war-effort when they assigned our missions; we had to follow the plan the best we could.

The commander as a result was in a tough spot. He could cover for the crew, and take a risk that something bad could happen. He could tell headquarters that we were down one crew and that would limit the amount of missions that could be accomplished. Or he could break up two crews to mix with each other and balance out the skill levels.

The third option was executed, the commander saying that at the halfway point during the deployment another assessment would be made and perhaps both crews could return to their original configuration.

Unfortunately, this was a pretty tough blow for morale for some, so early in the deployment. On the "office" side of the house, we had to give up some of the aircrew members we had gotten used to and created an efficient rapport with. Meanwhile, on the 'house' side of the house, we had to break up the living accommodations we had. Crews bunked together for crew rest reasons, which meant that we couldn't have people living together when they were on two different crews. Each crew operated at different hours which guaranteed one of the roommates would turn on the lights at the worst possible time for their sleeping counterpart.

It of course seems like a benign event and change. We found a solution to a problem, and all that we needed to do was have a couple of people move from one bed to another and work with a few other folks. This is of no concern, but for some reason, it added up in the minds of some.

The trials and challenges encountered up to this point of the deployment had been handled together as a team. Our same team had travelled from Clearwater to the deserts of the Middle East and mountains of Afghanistan. We had been with each other through the difficulties of leaving our loved ones behind, or the idiocy of forgetting an oxygen mask essential for flight operations. On the days we got less than adequate sleep, we frustrated each other with impatience yet still quietly covered each other's mistakes and rectified them without ever telling anyone. Those processes had been a necessary part of the deployment. We understood each other's limitations. To know these meant we could navigate the labyrinth of human sentiments and feelings – somedays, this person may say he has a migraine which is a clear indication to steer clear of him. On another, this gentlemen might miss his significant other which meant we should probably all go to the cafeteria together to socialize. All while someone else might want to over-socialize and spend hours on Instagram instead of sleeping before a flight.

The processes of group development were virtually complete. Out of the forming, storming, norming and performing steps, we were performing. Now we had to reset part of that process.

This reset was necessary, though. Like our crew covered each other's checklist procedures and watched each other's backs during tough descents into LZs in bad weather or during mortar attacks, our unit as a whole was a team as well.

And as a team, we would help each other out, even if it meant breaking up crews to create new, potentially even more effective ones.

Questions as to why our crew was the one chosen to fuse with the other crew lingered for some time but in the end, each of our crews were chosen for high-end missions over some other teams. After a serious phase of "storming" and dealing with the shortfalls of various positions and characters, life returned to normal. In fact, parts of the crew ended up developing new bonds and it definitely helped the crew by dividing it. This contradiction is explained that the six members ended up bonding in pairs. Each pair could lean on each other and with these stable, friendships being formed, we were much more able to be patient with each other and get along. By knowing we had a friend, the feeling of isolation that resulted from the shift in personnel wasn't so personal and we were able to develop as a team.

The initial shock and disappointment led to understanding and ultimately gratitude. The experiences learned from various crew-members showed how effective our military could be, having crew-members from a different unit each and still getting the job done well. Of course, the personal relations and life-experiences shared cannot be discounted and the growth from such interactions are an outcome we all can be grateful for.

As suggested by the squadron leadership, at the halfway mark of the deployment, the crews were swapped back. For privacy reasons and operational security, the aircraft positions and personnel that were swapped will not be mentioned.

## Triple Shuttles

The crew had been flying combat missions for a few weeks now, and we were now comfortable with taking on any mission.

Our contributions to the war effort were building up, but where the C-130 really was utilized was to access tiny, far-off and remote FOBs. These small outposts of allied forces relied on our fleet to deliver supplies and personnel, while also extracting personnel returning to the US or evacuating wounded troops. These places probably didn't have paved runways, and instead just said 'here's a small, kind-of-level patch of land: make it happen.'

So we did.

We normally could fit about three out-and-backs to these FOBs in a single duty day. This is based on the hub-and-spoke idea of connecting the Afghan villages and cities. We would fly from the Hub to a FOB, back to the Hub, and repeat with other FOBs. If we were connecting FOBs to FOBs, the number we could fit in a day could vary based on flight time etc. But the most common was the hub-and-spoke model of 3 in a day, which was known as the Triple Shuttle, or the Triple S. The S stood for multiple things; Shuttle, or the first letter of each destination, which were named: Sabarat, Stella and Sylvereal.

These S's were three Landing Zones that were accessible only to the C-130s, and as a result they were often on the mission schedule.

Our first flight of the day was to Sylvereal – a tiny FOB that was nestled against a mountain range, and saw a steady flow of insurgent terrorists that would strike before taking refuge in Pakistan. The small landing zone wasn't even paved, instead just being a collection of rocks strewn into a (sorta) straight line going up a small hill onto which we would land our multitoned warplane. Propeller warplane, that is. Filled with the food and supplies those suckers needed.

Regardless of what was behind flight-station 245 in the aircraft's cargo hold, the crew treated this outing like every other one they had: with the confident skills that permitted us to work rapidly without missing things. We could tell when something was out of place.

We activated the engines and then, like everyone does with their car, backed our airplane out of the parking spot and drove down the taxiway to the runway. The crew prepared the aircraft for combat, but didn't need to change anything on the airplane even though we were flying into dusty, rocky gravel roads or grassy fields. This fat propeller airplane symbolized freedom – it could land on a paved, smooth runway as comfortably as on those war-torn and abandoned dirt football (soccer) fields.

We took-off and proceeded to the first LZ of the day.

Flying through the clouds – never ending whiteness. It's an odd sensation to look outside and have absolutely no idea of where we are and where we are going. Frequent flyer passengers may often get a similar feeling: they look outside their oval window even though they know there's nothing to see as the plane just flies through an eternal bright cloud.

Unlike civilian airliners, though, mountain peaks were poking into the clouds, which meant we couldn't fly below the clouds (or we'd end up subterranean). Instead, we used our radar and altimeter systems to get lower and crest over the mountain peaks. Then we would dive down to the LZ, before the other obstacle.

Sylvereal was a small, dirt LZ close to the border with Pakistan. It resided within a few miles of a C-shaped mountain range with jagged peaks that stood 10,000 feet higher than the village we were landing in, just a few miles away. The other obstacle for our approach into the LZ was the Pakistani border, 9 miles away, opposite of the mountain peaks.

As we flew over the mountains bordering the small village, sometimes clouds would be blocked off by the mountains and amass over the LZ. We'd have to use our radar and internal systems to make sure we didn't run into the ground as we descended rapidly, also making sure we didn't stray into Pakistani airspace.

Looking outside the window and seeing absolutely nothing, not even the tip of the aircraft nose in the dense clouds, the radar sweeps were our lifeline, along with our internal navigation sensors to make sure we were headed in the right direction.

As we approached, the locals would hear us and suddenly we could be seen diving down, leveling off 1,000 feet. We'd be flying over the shacks and small huts and houses of the village. A river allowed some green grass to grow along the banks in an otherwise desolate world. We would keep our speed high from the dive until a pre-calculated point prior to the landing zone.

At that time, the pilot would pull back the throttles and the plane would suddenly slow down. Alarms would start ringing inside the airplane, the systems worrying that we were going to slow down so much we were going to fall out of the sky; if the landing gear is lowered, then the alarm doesn't sound. But we were still going to too fast to lower the gear.

When we reached the requisite speed, we extended flaps and then dropped the landing gear. We saw the LZ rise. Seeing a runway slope upwards was strange. But this makeshift patch of dirt we were landing on slopped upwards from our landing point, which would help us slow down, but first we had to get down on the ground.

N: 2.5 miles to go, start the descent, on glideslope.
CP: 125, down on 3.
E: Speeds are 121, 111, 100.
CP: Checks.
N: Half a mile, 200 feet… 100 feet… 50…40…30…20…10…

We touched the dirt and immediately slammed on the brakes. We were going to use every inch of this tiny patch of land to stop our massive airplane.

When we got to the end of the runway and were travelling at a walking pace, the plane started to turn around as we opened the ramp and door, keeping the engines running, ready to leave at a moment's notice.

E: I see people on the roofs.

Indeed, plenty of men were observing us from the roofs flanking the small 150 foot strip of rocks and dirt we landed on.

L: 12 pax, 3 pallets, 5,000 pounds, reported the loadmaster as he continued to work regardless of who was watching.

The front end crew immediately started the calculations for take-off, including the speeds and distance required while factoring the down-sloping LZ.

L: Ramp and door, cleared to close.

That was the signal that the plane was ready to go.

We ran quickly through the clean-up actions to make sure we were ready to take-off and soon after, the throttles were pushed to their maximum power while holding onto the brake pedals. The ceramic brakes held back the 4 engines generating 4,350 shaft horsepower each, ready to perform a maximum effort take-off.

After a few seconds of racing down the sloped LZ, the plane hopped into the air, leveling off a little after the 50 foot bounce up. We accelerated fast and low over the Afghan village. As we approached the mountain range, we started to climb again, immediately finding ourselves in the thick, white clouds and losing all visual cues of the world around us. We were entirely reliant on our instruments again.

Using the radars, attitude director indicator (ADI) and good old maps, we navigated ourselves to a higher altitude before crossing over the unseen mountain tops. Once that was accomplished, we contacted Tetra to tell them we were out of Sylvereal, close to the border and were headed back to The Hub, which they approved without any hesitation.

Forty minutes later, we were back at the busy Hub, where we were also busy. We were offloading our passengers and cargo. The fuel truck coming to give us a sip of gas was late, which was a little frustrating for the other sorties down the line, but we figured we could easily catch up the time with some savvy loading and unloading at the next destinations. We weren't expecting to onload a lot of materials, but the Operations Officer on duty walked to the airplane with a large box of 'morale' supplies: peanut butter, chocolate, candy and other American products that were difficult to access at the tiny FOBs we were resupplying. Embedded with these small consumables were letters from various schools and volunteers across the United States with warm messages of support.

With the valuable package tied down in the back of our large cargo bay, the plane raced back down the Hub's long runway and raced toward our next stop: Sabarat.

As before and just like any ordinary flight, we got to Sabarat. As we dove straight down from our cruising altitude, we weaved around a flight of helicopters that had been cruising below. We didn't do it to be cool – we just hadn't seen them. It's a good thing the Traffic Collision Avoidance System (TCAS) worked and yelled at us to avoid them, because otherwise that would have been the end of this story.

But as usual we landed without any issues and immediately loaded up our cargo with the engines still running.

Once completed, we began the same preparations as before. We calculated our power requirements, distance available, the weight of cargo, hot temperatures, high altitude (hot, heavy and high were the three H's that were never convenient and limited our performance). With the data calculated, we determined our procedures like we had before. We would calculate a time where if we were past a proper speed then we'd take off before the end of the landing zone. This was good: it meant we wouldn't fall into a chasm or crash into someone's house and explode into a memorable but ultimately fatal fireball of death.

P: Time for 11 seconds. Ready, ready: hack!

The plane's brakes were released and the plane started to instantly roll down the dusty strip. Eleven seconds later...

N: Time!
CP: Reject!

The crew stood on the pedals and threw the throttles into hard reverse. All of our heads and arms flew forward, our bodies held back and crushed into the harnesses of our seats. Our speed was slowly decreasing at first, creating some worries that we weren't slowing down fast enough.

Fortunately, the speed needles continued to slow down and at a faster rate as we reached the end of the landing zone.

When we came to a full stop, we all breathed a sigh of relief. It wasn't only because we weren't going to die but also because we were not longer being flattened into our harnesses. A tip of the hat to the loadmasters and the aircraft engineers – the load had not shifted at all during the emergency all-stop, even though there had been a lot of inertia. The multi-hundred pound pallets and bags had stayed put, putting incredible strain on the locks and tie-downs, but ultimately not budging a single millimeter.

We taxied back to the starting line again (we couldn't just turn around where we stopped and take-off the other direction: there were aerostats and towers right at the edge of the runway – that's just no good.) While we slowly travelled back up, using our engines forward and reverse thrust to control our speed and minimize the use of our very warm brakes, we talked about what happened. The speed read on the dial was not what we were looking for, so we rechecked the data and updated the information for the next attempt.

The throttles were again pushed to max while holding the brakes and when the power was set, we again released the brakes and raced down the narrow runway patch.

Airborne! This time there were no issues and we successfully lifted our fat, heavy airplane off the ground using our futuristic propeller technology.

Fortunately, after another 40 minutes, we returned to Hub. Unfortunately, we were running late – not only because of the rejected take-off. We had just had a slow turnaround time before hand and were getting late. Being late is never good. When you're a member of the military, it's even worse. When you're in a warzone, you're just downright rude.

After swapping out materials and passengers again, we began our departure movements. Our next destination, Stella, was manned by a small Special Forces team, and an Air Force Tactical Air Control Party (TACP) member assigned to the SOF team would serve as our air traffic controller, with nothing but his battle gear and a handheld radio to talk to us. There were a dozen civilians that had to be picked up from the area, and we had to resupply the Special Forces team with ammunition and other explosives before they moved on, and before we ran out of time based on our own flight regulations – after all, this story still had to be about us.

We first had to travel to what was called the Haz-ramp, a parking spot at the airfield that was far from pretty much everything, because it was meant to load hazardous cargo (like bombs and other munitions). Since it's so far, taxiing there requires some time, which was slightly frustrating given our current time-keeping abilities.

Once we loaded up the hazardous cargo, we began our long travel back to the proper end of the runway to keep the timeline and make it to Stella before the TACP left his sniper's perch. But we still didn't have clearance from HQ. It didn't matter, we maneuvered the plane off to the side of the taxi-way just at the entrance of the runway, where we would wait while not being in the way of other aircraft, fully expecting clearance from Headquarters in Qatar to come down soon. They controlled our PPR though, which was the permission for us to land. They may have already decided we were too late and cancel the mission. So we waited.

N: We got Papa Lima.

P: Ok. Let's call tower and request an immediate. We gotta get going.

SHANCS 55: Tower, SHANCS 55, holding short of echo, request immediate take-off.

Tower: SHANCS 55, winds are zero-five-zero at 8, runway zero-three zulu, cleared for take-off.

SHANCS 55: Cleared for take-off runway 03-Z, SHANCS 55.

Tower: SHANCS 55, no delay, you've got a Beechcraft 5 mile final.

SHANCS 55: SHANCS 55, copies. No delay.

McKay blasted through the lineup checklist as we taxied onto the active runway, flipping switches and twisting knobs on the panels as he called out the checklist from memory.

E: Seat belt and shoulder harness already called fastened and unlocked around the board. Exterior lights?

CP: Set, copilot.

E: Set engineer. IFF/TCAS is set, line up checklist complete.

The throttles were immediately pushed to their take off power as we initiated a rolling take-off.

<u>E:</u> Power set!

Without even slowing down from our arrival onto the runway, we quickly took off and maneuvered to the right a little bit away from the runway to clearly show we were no longer a factor. Tower got the clue and called over the radio.

<u>Tower:</u> SPY 76, check gear, winds 0-5-0 at 8, cleared to land 03Z.
<u>SPY 76:</u> Cleared to land, 76.
<u>Tower:</u> SHANCS 55 switch to departure.
<u>SHANCS 55:</u> 55 to departure.

After talking to the appropriate agencies, we headed toward Stella quickly and about 25 minutes later, got on the radio. Our TACP member answered, perched in his small, shredded tower, with nothing but his rifle and an MRE he was nibbling on. We were going to be supplying him with more bullets and food for both him and his team.

Using a pair of binoculars, the Air Force special operations member controlled us and guided us down to the landing zone.

Once on the deck, we rolled out the plane to the end of the LZ. From behind abandoned containers that had been used long ago and looked rusted now, a group of people waited for us to open the ramp and door.

When we did, the loadmasters signaled to them to wait a few more moments.

"Time to re-rig," said the loadmaster.

Carter rushed to help the two loadmasters start putting up central pillars and attach canvased seats. The plane had been prepared to deliver the ammunition and other gear pallets, but now it had to be reconfigured to carry passengers. While the engines continued to roar outside, the three literally rolled up their sleeves and got to work, using their now well-honed habits to get the job done as quickly as possible. It would take the crew of three a little over ten minutes to furiously set up enough seats for the number of passengers expected.

When a rocket attack began.

"We need to take cover," said one of the aircraft crew-members. There was no way we could be ready to take-off soon, and our big airplane could easily be taken for a target – the only, giant noisy thing around for someone bombing things to aim for.

"There's a bunker right behind the jet."

"Shut it down," commented someone else.

The engines were immediately shut down while the area came under siege. One of the crewmembers gathered all the secret materials from the flight deck while two others pulled the handles to shut down the engines and secure the parking brake.

Rushing out of the airplane, the passengers and crew bolted toward the nearest reinforced concrete casting that would serve as our shelter until the mortar bombing attack ended.

While the area was being attacked by insurgents, we all sat patiently in the bunker, safe in the knowledge that there was absolutely nothing we could do to stop it. We were no longer in the airplane that was a huge target (and so far had not been hit by a mortar round), instead taking cover in a reinforced shelter. There was nothing to do but wait.

After a seemingly impossible wait, the all clear was given. The crew didn't wait an extra moment and rushed back to the plane. Having kept an eye on it throughout the attack, no mortars had hit around the plane so we didn't need to worry about battle damage.

Instead, we rushed in, flipped on all the switches required and immediately began to fire up the engines. One of the passengers asked why we hadn't just left the engines running instead of wasting time restarting the engines. One of the loadmasters politely told him to go die, while the other simply explained it to him: there's always an inherent risk that engines catch on fire (especially if it's being mortared), which means someone always needs to be available to pull the emergency fire handles inside the cockpit to shut off the motor and potentially activate the engines' internal fire extinguisher. Since no one was going to be left in the airplane, the engines had to be shut down.

Within minutes (certainly a shorter time than the attack lasted), we were restarting the engines. We only started two of the engines before we decided to start moving. We recalled the load inside and kept our procedures rolling along.

Once we started to taxi, the two other engines were activated while we continued moving.

Our years of training were paying off as we spouted off checklist items, and we were all conscious of what was critical for flight and what we could get away with not checking until after – when we'd be safe in the skies.

We slammed the throttles forward and immediately rolled down the small landing zone we had arrived on, ignoring the smoke emanating from where the attack had taken place, or the hive of activity resulting from the Quick Reaction Force...reacting.

As we lifted off the ground, we stayed low to build up speed and ran our checklists quickly, without a pause in speech, anticipating each other's responses:

E: Landing gear...

CP: Up

E: Flaps?

CP: Up

E: Landing ligh...

CP: Set.

E: AUX pump...

CP: Off.

E: After take-offs complete.

N: We've got the speed, start your climb.

We pulled 2 Gs as we shot higher into the sky and away from the ground, putting thousands of feet between us and the bombed base (and the surface of the Earth).

After we returned to The Hub, life continued normally for all of us. We laughed over a poor landing performed in pilot training, telling stories while the passengers were escorted away from our plane. We joked as we returned to the Ops building and logged another day flying between small FOBs and hazardous dirt strips, not thinking about the risks. Our biggest concern right now was figuring out what would be served at dinner, and what time it was in the United States to see if we could talk to our loved ones.

The only note of the attack was written in the flight log explaining why we were late:

"Got IDF'd at Stella. Shut down and evacuated to bunker w/ pax, which is why we were late. Sad face. All else, NSTR."

We would repeat this day many times. The small LZ were what the C-130 specialized in: getting in and out of small outposts. It was light enough compared to a massive C-17, had the range to reach far flung outposts compared to helicopters, and could defend itself if it needed to.

Triple-S was the nickname of only one of the routes we flew to the various LZs and villages around the region. That route had three LZs that had names which started with an S, but not all LZs made such naming easy or possible. Some were more scenic than others, but the risks and landing procedures were similar and most were in mountainous terrain. One was occupied by Australian forces (which we called Tee-Kay), while another (with a tiny paved runway) served as the mini-air base where Afghan Air Force students were learning to become pilots (which we called Dandyland). Another LZ was a favored target for insurgents and it sat exceedingly close to the border with Pakistan. There were many, many LZs to go to, which kept our missions challenging, dangerous, and in the process, interesting.

Other times, we would just have to pick a place to land. The radar on the aircraft could make-out telephone poles, which mean from high above the ground, we could make out the undulations of the terrain below, where a small team was waiting to be extracted. We'd find a road and say we would land on it, before an intersection, and that was the rendezvous point. These types of missions and abilities were reserved for the more secretive missions (that means we're not going to talk about them anymore) or for some emergency medevac missions.

Habits would develop as we flew into the various villages niched in the mountains of Afghanistan, which meant we were getting pretty good at it. That didn't mean there weren't risks. During the deployment, one of the missions saw the plane's antenna on the belly of the plane snag the Constantine-wire atop the T-wall surrounding the dirt strip, coming within 8 inches of hitting the wall and crashing. We were lucky.

Crews still face tiring, sometimes exhausting schedules. Stressors sometimes cause us to have poor sleep as well. Flying a C-130 is not like flying an airliner that has autopilot, especially not down to the small LZs in mountainous terrain of hostile Afghanistan. It can be a work-intensive process.

In 2016, many years after our crew had left Afghanistan, a C-130 went down at one of the FOBs we serviced, killing the 4 crew-members and 5 allied troops being transported, as well as a team of FAST security forces that were there to guard the plane while on the ground. Additionally, 3 Afghans were killed when the airplane impacted the ground and hit the building they were working in.

For our crew, even with those dangers, weather, human factors, enemy activity, we continued to return to the LZs. Those were the risks we faced every time we flew into them until we evacuated them and ultimately closed them.

Indeed, the crew would soon find itself pulling forces out of some areas.

## No Action

After a month on station, daily life in a combat zone began to feel like living forever in a day. Without the anticipation of weekends or days off, each day began and ended in the same fashion. Morning and evening, mortar attacks from nearby Taliban fighters were common. The daily explosions coupled with the weight of the Beretta M9 pistol strapped to our chests or legs reminded us that we were firmly entrenched in a combat zone. Of course, our flying days were the opportunity needed to break up the monotony and leave the confines of our military fortress, although flying a triple-S mission sometimes got boring, especially when the weather was nice. But when inside this barbed-wire oasis filled with Americans, life became very repetitive.

Because of the water-fall like schedule created from our combat flights, the crew lived in a constantly changing timeline. A normal day could easily start at 11:30 PM – the time most normal humans would consider bedtime. But for most of the crew, that would be the time we would wake up and proceed to the communal showers and bathrooms to prepare for the upcoming day of rest.

It's still difficult to say if it was simply out of habit, or perhaps loneliness, why most of the crew would gather together at the start of the day. After our showers, we would all proceed to the DFAC, which was serving Midnight Chow; the last (or first) hot meal of the day. Breakfast, consequently, consisted of dinner food, unless we grabbed packaged cereal from the snack tables in the cafeteria to smuggle home. But a plate of pasta and chicken was perfectly fine for most of us as we proceeded down the line, wearing the exact same uniform we had worn outside of our rooms every single day of our lives in the nation we were occupying, at midnight.

After eating quietly and watching the few television screens hung on the wooden beams of the cafeteria presenting AFN information, the crew gathered the continuously similar cardboard plates and threw them out into the trash bins at the exits of the DFAC. We then threw out the soda cans or the juice-boxes in separate trash bins. There were no glasses or utensils to place into a container or conveyer belt to be cleaned for later use. Every day, all of our food was eaten from cans and cardboard plates. At least it made it easy to steal some cans of soda and put them in our cargo pockets to then consume in our dorm rooms.

Following the usual morning dinner, we proceeded to the Operations building. If anything interesting was happening, it was at Ops. While our crew had a day of rest, our alternate crew was just getting ready to fly. They had their day off yesterday, while we were flying. But this was around 00:30 / 1 AM, which was also prime-time for an insurgent attack. Rocket and mortar attacks were fairly common, and the biggest concern after months of being bombed was not so much being hit – since there wasn't much we could do except take cover, but it was where we would take cover. If we were lucky, the attack would happen while we were in the DFAC, and we would simply all duck underneath the tables, but still have our cookies to nibble on and televisions to watch. If we were outside though, we would have to dive to the ground during the initial attack, and then rush to the nearest bunker. Inside the bunker, we would wait – where there was no climate control. But at least we could hear the helicopters and Quick Reaction Forces launch to retaliate and re-secure the base after the attack. We could 'follow' the progress of the operations. Usually after about 10 or 15 minutes, the attack would be over and we would continue our travel to Ops.

There had been one incident where a bombing started while two of the crew were walking to the ops building. They took cover inside a bunker that was within *sight* of Ops (which had Wi-Fi, contrary to the bunker). After waiting a few minutes after the last impact was heard, but still hearing the loud "Incoming" calls over the loudspeakers, the two crewmembers rushed to the Ops building to wait out the attack there. Unfortunately, as they rushed in, they ran right into the DO. The second-in-command of the unit was not very happy to see them. Clearly caught breaking the rules, the two crewmembers were instructed to send out a unit-wide email reminding everyone what to do during an attack to ensure everyone's survival.

Inside of the Operations building, the crew all did the same thing before breaking off to different tasks: check the mail corner. In the back of the Ops building was a shabby closet inside of which our mail, delivered by a member of the unit, would be waiting daily. Packages and letters from home would often fill the closet, and the unpredictability of the postal system meant it was always exciting to receive something. While email and video chat were the most common means to communicate with those back in the United States, physical mail still remained one of the most emotionally rewarding events of the day. Inside those letters or packages were handwritten notes, along with actual goods from the United States; be it a box of cookies or the extra pillow you forgot. The items themselves were often worthless on a monetary level. It was the thought, the emotional connection to the person who sent it, that was truly significant. The fact that someone thought of you while you are thousands of miles away, and took the time to send you a package is what, subconsciously, created such an encouraging reaction. Boxes were ripped open, and quite often, the contents shared among the crew. Carter would receive a specific type of chocolate that Telford also enjoyed, while Hammond received Nutella; this allowed Carter to declare that he could stay longer in Afghanistan so long as he had a steady supply of Nutella. Of course, personal orders from Amazon, such as mattress pads or extra razors for shaving, were harder to share but just as enthusiastically received – it meant that one was about to have their local comforts evolve for the better.

After the sharing of these prized packages among the crew, everyone split off. Some went upstairs into a makeshift movie room to watch DVDs handed down over rotations of deployments, while others checked the next day's schedule. Many of us, though, proceeded to the three government computers located in a small room that were for anyone's use. These computers had the required card-readers and digital certificates needed to access our military emails. Our days off were not only for us to rest, but also designed to allow us to focus on other responsibilities expected from us. Email was the lifeline for everyone to home-unit. There were often different types of training courses that needed to be accomplished, or documents to be signed and emailed back to the United States for the next day's duty day. The crew could easily notice when it was the weekend in the United States, since emails were slower to leak through to us. But we still had other military duties to focus on, and now was the time to concentrate on them.

Today's main point of emphasis for the officers of the crew were going to be Letters of Evaluation. Being deployed in the warzone was nothing special, particularly a decade after the September 11 attacks that motivated such a war. Everyone on the crew was looking at what came next, beyond the deployment. How to further progress their careers and ensure success. Out of the six lines from a military officer's performance report, deployment achievements filled one, maybe two lines. The rest was dedicated to officer performance in other areas – often things that are unrelated to their jobs or skills as military officers. So to further careers, the officers of the crew wrote tweet-sized sentences of praise regarding the deployed accomplishments of their subordinate crewmembers.

Fortunately, there were a few to write home about: combat airdrops were still significant exploits given the risks and complexities. Different strategic missions were performed as well, and when something went wrong, the skills of crewmembers were what allowed everyone to land alive, and avoid media attention. So, occasionally, the performance reports were rewritten, polished or refined to reflect the successes. Thesauruses were used and a massive amount of statistics were compiled in order to quantify the contributions (small) to the military operations taking place (more in Chapter Amusing Details). But in the end, the investment in both time, treasure and emotion to train combat aircrews for the war they were currently fighting would be recouped through those single lines edited to look like:

- Conducted NVG combat airdrops; dlvr'd XX bundles/XX tns cgo in hostile airspace—critical CJSOTF objs supt'd

The rules for these bullets are simple: you are given 110 characters to describe an event, how you did it and the impact. You cannot leave any white space at the end, but you can't go over the 110 characters into another line. This means that while the idea of the bullet point is usually decided early on, any change in the words can impact the rest of the bullet. In the example above, the impact was that the team was able to support special operations objectives. 'Spec Ops' took more space than 'SOF'. So 'SOF' was used, and then 'objectives' could be typed out... Nope. The '-es' of 'objectives' didn't fit. So instead, 'objectives' was shortened to 'objs' and SOF was expanded to CJSOTF.

The bullet can consequently be 'translated' to the following:

- Conducted night-vision combat airdrops; delivered XX bundles equating to XX tons of cargo in hostile airspace—critical Combined Joint Special Operations Task Force objectives supported.

After finishing up with our Air Force responsibilities back in the United States, the crew would waddle back to the dorms. It was around 3 AM by then. We could still hear planes take off and land, their bright lights and strobes glowing as the planes raced down the runway to gain speed. When they took off in the middle of the night, they would extinguish those lights to instantly vanish into the dark night. The only indication of their existence remaining was the fading roar of the jet engines.

Back in the dorms, the crew would break up. Some sat at their makeshift desks, made from wooden remnants of old demolished huts, and work on responding to personal emails, while others would catch up on their current TV shows. Deployments were perfect times to focus on education, going to the gym and getting fit, but mainly to binge watch television shows. A black-market system of TV show trading was instantly created when the crews arrive – hard drives filled with movies and television shows were traded in order to boosts each other's libraries. Old shows such as The West Wing all the way to the most current season of The Walking Dead or Game of Thrones were available to watch. Local merchants also sold bootleg copies of movies which ensured that our entertainment would never run short. Different crews also networked each other's laptops to play different video games, and some deployers are brave enough to bring an X-Box or PlayStation with them to plug into a small television ordered from Amazon and delivered via military airlift (which we were fortunately a part of and understood very well).

All of us would watch some episode of a television show for entertainment until we got bored. Then it would be time to go to the gym. And this was the only time we were permitted to walk outdoors without our guns strapped to our legs or chest. It was a strange sensation to be walking from our containers through our gravel paths to the large tent that housed the gym, with its weights, floor mats and fridges filled with bottled water. The base was very quiet at 3 or 4 in the morning. Armored HUMVEES and MRAPs would often drive down the base's primary road, but otherwise, the walk to the Clamshell (the name of the tent-gym) was quiet. Fortunately, so was the gym. It was cool outside, which was also a welcomed change to the usual heat and dust. If we forgot our bottle of water for our walk from the dorms to the gym, we didn't feel like we could die of thirst.

Returning from the gym always meant a shower. Like most dormitories, we only had communal showers and bathrooms on the floor. Similarly, we also knew which shower was more reliable than another – which one could provide us with hot water, and which one had water pressure. You couldn't have both – if you wanted to call yourself a warrior, you could not have both. The air was usually stuffy and humid because of the poor air circulation. But after being at the gym, it didn't matter. The dust that was outside was already sticking to our skin and our damp physical training uniforms, so the faster we could shower, the faster we could change to our equally dusty camo-uniforms.

While we would be in the middle of our day though, the person in the next shower stall could be ending their day, while the other gentlemen trimming his deployment mustache over one of the two sinks was just getting his day started. Our little community lived such a staggered lifestyle that all we did all the time was say 'hey' – certainly not 'good morning' – you never knew if it was their morning or if they were having a good morning.

After the shower, between spending time at the gym and in our rooms, all of the crew somehow got back together within the two hours since we had returned from Ops. It was at that time that we would all agree to grab some lunch. Of course, at 5 AM, lunch was to consist of French-toast sticks, scrambled eggs and sausage – again. But the DFAC was more crowded now. Much of the base was preparing for their day of work. That meant that different offices would be open for us to visit as well, which included a building nicknamed the Pink Palace.

After our breakfast-food lunch, we headed to the Pink Palace compound in order to solve some issues that remained with our finances. Some of us had to acquire a new Eagle Cash Card, a debit card used for almost transactions at FOBs, since carrying cash or personal credit cards was asking for trouble. The Finance Office would issue a card that you could load with funds from your bank account directly (like a prepaid gift card). Others of us would sign paperwork for different pay benefits we were entitled through previous missions or the current deployment, while others would be investing their money. The offices in the Pink Palace also issued us different ID cards and certificates that would allow us to move freely across the base. Once we were all done with what had to be accomplished, we took the opportunity to stay out for a little longer.

As the Sun would slowly rise, signaling the start of the day for most people, the few stores were also opening. This meant that for us, it was time to perform what colloquially was known as a shelf-check: a wandering to the small base shopping store to see if anything interesting had finally made it to their shelves. It would give us the opportunity to go for a walk and indulge in our American consumerist needs. The Hub's Base Exchange, or BX, was located in an old converted Soviet hangar, and it was generally stocked with a few magazines and some food products from the United States. There were Pringles, Gatorade bottles, cases of Mountain Dew that we could buy, but since all food was free at the DFAC, we rarely bought any food or drinks. The soft-drinks for sale were directly imported from the United States, which meant they were made differently from the locally produced Coke products (American drinks are made with high fructose corn syrup, while the rest of the planet drinks Coca-Cola with real sugar), so unless you wanted a dose of high-fructose corn syrup, very few of us bought the drinks. What was often purchased were razor blade replacements or some extra shampoo. But the shelf-check in the store lasted only five solid minutes before we would turn around and start to wander back to the dorms.

On the way, we would pass a small bazaar manned by locals and where they sold bootleg everything. From knives to DVDs, you could watch the latest movie that was just released in cinemas across the United States in the comfort of your own dorm room. Seasons of television series, copied from Netflix, were also available. Knives, Afghan dress-suites, and "quality" Rolex watches were also available. But after checking the products once, we had lost interest. So we continued on, passing the small Pizza Hut that tried to serve a semblance of a pizza (with difficulty though, given the environment they were operating in).

The crew did often decide to indulge in a Green Bean – Honor First, Coffee Second, cup of coffee or fruit smoothie. The closest thing to a Seattle-style café, the sole-deployed coffee shop was a typical stop for us. With our drinks in hand, we would then wander back to the dorms.

We would often take a small gravel alleyways between rusting containers and arched tents instead of the single major roadway. This was done in order to minimize passing by fellow service-members, which meant fewer opportunities for us to have to salute each other. With coffees in hand, it was a noble goal.

Before getting back to our containers, we stopped by the laundry tent. Unlike our homes in the United States, to have our laundry done, we would drop off a bag filled with our dirty laundry at a drop off point. On the side of the bag was our name and the last four digits of our social security number. Three days (or so) later, the laundry returned, clean and folded. The challenge was picking it up, and the key to a smooth and quick pick up was a laundry bag that stood out. Otherwise, if it was a white or meshed bag, they all looked alike and when it came time to pick up yours in the multitude of other bags, then it would just take forever. Carter found a trick by defiantly carrying a bright pink bag (making himself an easy target to any sort of attack; mockery or otherwise), while others had duck patterns on their bag.

America's fearsome warriors soon returned to our dorm cell, armed with coffees and laundry bags (and 9mm side-arms), where we changed into our comfortable pajamas again.

We humbly returned to our films and television shows, video games, snacks and care-packages. Sealed in our rooms, we would simply and quietly keep to ourselves. Ignoring the problems of our world for one day, was actually relatively helpful. We faced those problems every day when we flew in combat, either facing off with the challenges of the war, or hearing other things impacting our ability to help in this conflict.

Other times, the hours spent with our roommate were time to vent frustrations and escape from the constant reminders of work and war around us.

These were the moments when a real human connection was made in addition to the actual operations. Discussions about politics, economics, or even social ambitions such as having children were discussed.

Regarding work, the concerns that many have in their civilian work life, about how to impress the boss, prevent someone from taking credit for an idea, or ensuring the proper climb in the company, also exist in the military and in a deployed environment.

Because of the small community and the lack of opportunities to leave work behind, it would sometimes lead to psychological exhaustion. And when someone is exhausted, they become frustrated which causes them to be even more tired, creating a dangerous, cyclical movement. The crew members, behind closed doors, would sometimes become cynical about the way the crews were managed, or about how the personnel were viewed within the unit.

There would be weeks where things would slowly get more and more depressing. And then another set of weeks where morale would be sky high - we were proud of our work, enjoying the way our private lives were going and maintain great social connections with the United States.

During the down parts, the primary source of irritation would be a perception that the individual was not valued. Even while we were taking part in the war effort, logging combat hours every other day, flying multimillion dollar airplanes, even delivering humanitarian supplies to regions devastated by natural disasters, our small island disconnected from the world and the normalcy of these events made us jaded and misperceive our privileges.

In the relative world of our combat airlift team though, we still vied for various honors, and more importantly, for a good reputation. The worries were borne purely out of differences of personality and were sourced only because of the human condition. Some people like others, which implies that it is possible to also dislike someone.

Members of the crew could sometimes dislike someone based on their skewed perceptions. This was the consequence of placing highly trained, expert aviators in close proximity to each other and 'competing' for attention at the higher levels. Behaviors from a leader to a subordinate would be frustrating if someone didn't like the rules they were following with the people they were stuck with. They preferred different leadership styles, maybe.

But behind the closed doors of our metal container rooms, we were in a safe zone, where each pair could talk and vent. We could do the same as a crew from time to time. Having each other to talk to was crucial to ensure we stayed generally positive throughout our deployment. Friends and family back in the United States were also vitally important, especially to disconnect from our work and military lifestyle. However it was also this disconnect that they had with the military that sometimes made it difficult for them to relate.

This lack of ability of relate sometimes created a different set of problems.

But again, these problems were limited.

Most of the time, depending on the time difference, many of us would call back to the United States to wake up our loved-ones and tell them we were still ok. That was the most important. It seemed trivial to those deployed at first that they had to call to check in like children going out with friends, but not calling home would often put loved ones under a lot of stress. They would wonder what happened – why aren't they calling? The news would never really report anything positive out of Afghanistan. So we called home, and we would laugh on video chats or over the phone. We would get to live life vicariously through those that remained in the United States.

These recaps from the United States, home, were often another source of conversation for our container-room dialogues. Profound considerations about our futures, our pasts and hopes were discussed.

Roommates would learn about what had happened back home while we had been flying and then sleeping. For some, a dog was not going well and would most likely be gone by the time we returned from the war. For others, it was the news that their loved one was being promoted. In a rare case for crews, no one on our crew had children. So we didn't have to worry about the bad grades some had in school or any other parental issues that are sometimes tough for single parents. But that didn't mean the topic of children did not come up in conversation. Some considered founding a family, which led to conversations about our own families and memories we held of growing up – life before the military. Or perhaps even our motivations to sign up in the armed forces.

Other times, one of us would go sit in the hallway, using the lawn chairs we brought over with us. This would allow our roommates to continue sleeping if we had woken up early, or talk to family on the phone privately. We wouldn't be in the hall to socialize, though. Even though we were awake, that did not imply that other crews were awake – quite the contrary, they would be sleeping. Instead, we would go for our moments of privacy: staring at the glow of our cell phones in the often dark halls (they were kept purposefully dark to prevent light from pouring into our rooms when we opened the doors), we would quietly surf the web, look at the stock market to see how we could make it big, or research houses we dreamed to live in.

Another method to speak without bothering others was to go into the staircase of the containers, or walk to a small 'morale hut' where a Wi-Fi router had been installed. But like everywhere else, including the room, others would transit through the staircase or also be at the common 'morale hut', or the roommate would reenter the room. This was more frustrating for those at home than for those in Afghanistan – we understood what life was like at The

Hub, but explaining it to someone in the United States was a little more complicated.

During certain weeks, the crew would be expected to conduct additional ground duties, including being a liaison to the higher levels of base leadership with tower. That meant that one of the officers from the crew would have to actually sit in the tower for about 8 hours and work with the air traffic controllers.

At the end of the shift, someone from the crew would grab one of the parked pickup trucks (massive F-350 trucks) and drive around the base to reach the tower. It was one of the few times we could drive. The US broadcast AFN (Armed Forces Network) radio stations which meant we could keep in touch with the newest 'hits' back in the US. Country music would sometimes be playing as well, and there are few things that bring out a sense of being 'American' like listening to country music (in a country we were occupying). Those drives to go pick each other up at the tower were another time where we could be alone in our own world. We would just be driving a car (tank-sized pickup) around the long roads that circled around the runway of the base. SDO duty was at night, so we always drove to the tower at night, where the world outside the base was lit up by massive flood lights, but inside the base, it was all dark except for whatever could be illuminated by our headlights.

Of course, the flight line was always lit, and because Ops and the parking of our small fleet of trucks was on the flight line, we always had to bring the vehicle back onto the secure area. After picking up our crewmate from the tower (sometimes bringing them a warm meal from the DFAC), we would get to the Entry Control Point to the flight line. There, Security Forces would check our vehicle in detail – we had to present papers, our own badges, open the hood of the vehicle, leave our doors open as they did their sweep. When it was done, they would lower the massive, hydraulically-actuated gate and we could drive along (after having waited for other vehicles ahead of us to do the same. That's usually when the two crewmembers would catch up on what had happened over the shift at tower.

Even after that thorough inspection, though, the driver still had to be careful – we were now driving on an active flight line. Just like in normal airports in the United States, there were crossing intersections between aircraft and cars or trucks. The difference here was that the T-walls protecting the airplane would block our view of most things going on so we had to be extra cautious. If an airplane did cross our path or faced us, we would have to turn off our headlights to avoid blinding the aircrew. Helicopters would hover low over the ground toward their taxiways of departure. It was exciting to think about, but in the act, it was like any other day at the office. Our blasé attitude was indifferent to the amazing symphony of activities taking place around us, from medevac helicopters bringing in wounded while a small spy plane took off to find those responsible. Being surrounded by these activities and the stimulation and perils all the time were what reset our threshold for excitement and danger, and what influenced our desires.

It's odd to yearn to leave our little corner of the world. We live the same way when we're home: we get up at home, go to work, go to the gym, go to the supermarket or restaurant, and then we go home again. It's the exact same thing in Afghanistan. But somehow, we instead believed we were trapped and sometimes, however rarely, that we were not doing anything of value because we were so used to performing activities that others would consider amazing, unique or exceptional. Thinking that we could not leave the confines of the base for months on end could have a psychological effect, especially on the non-flyers. Unlike the combat aircrews that had the opportunity to leave the base and visit the world, traveling to various countries outside of Afghanistan, the support crews were locked in place at The Hub. Unfortunately, with every aircraft departure and recovery, they were reminded that they were trapped.

## Operational Challenges

Being deployed is generally not considered to be a simple walk in the park. Simply adding to the stresses of being far from home were mixed in the rocket attacks launched against us, the risks of flying in such a complex airspace, and the overall tempo of work, where there were no vacations or weekends to look forward to until we would finally return home. And these stressors did sometimes reveal themselves through people and their interaction amongst each other.

Many missions were often signed off with four letters: NSTR – Nothing Significant to Report. The missions were accomplished and that was it. That's the way we always wanted to keep it. If we deviated from NSTR, then that usually meant someone was scrutinizing our work. And rarely was it for something we were doing well. But in such a high paced environment, with almost 24 sorties a day, there were bound to be some challenges to maintain our anonymity from scrutinizers, sometimes even making mistakes that resulted in some raised eyebrows.

On a mission from The Hub to the British Isle, Tetra asked us to fly to our destination by flying over two different sites before proceeding. This was to avoid flying through airspace that was being used by strike forces; gunships and fighters.

When we received the instructions, all the on-board navigation computers completely failed.

We no longer had anything to navigate off of, and we were blind as to where the busy airspace that was blocked off was. Remembering the last radio call, the crew pulled out charts and went into pure improvisation mode, reaching back to our most basic navigation training. Using intermediate points on the ground and a variety of navigational aids, we managed to avoid the airspaces requested by using our fingers as rulers, and approximating our position by quickly calculating the distance we travelled at our current speeds. Using the intermediate points, we finally made it to a point where we could see the British Isle. Given the clear weather, we were able to visually continue our way to the base and land without incident.

On the ground, we tried to restart our systems but they would not come back. To complete the rest of the mission, we simply requested to fly on instrument airways – the air-roads that airlines usually fly. We would be unable to accept any unexpected taskers, and we didn't even talk to Tetra. Instead, we would talk to normal controllers like any other airliner, telling them where we were going, and because of the mountainous terrain in the nation, where we were at certain waypoints since they couldn't detect us using radar (the mountains blocked the signals). Someone in a control center was guestimating where we were to ensure other planes wouldn't run into us.

Fortunately, this loss of systems had only occurred once and we were able to still wiggle our way through the mission, accomplishing the mission without needing to say anything else other than NSTR.

We returned the plane to the maintenance team, and told them about the issue. Only maintenance ever knew about the problem.

Unfortunately, another event caused the crew to be reprimanded by the commander after the aircraft sustained damage. If it was battle-damage, the commander didn't talk to us. If it was our fault and wasn't battle damage, then he did talk to us.

For this time, he did talk to us.

Which meant we had damaged the airplane by accident.

It was a simple accident, where an item was left underneath the ramp of the aircraft as the ramp was lowered. The hydraulic pumps opening the ramp continued to open the ramp even against that item that was underneath, creasing a part of the aircraft's metal skin. The loadmasters quickly noticed the problem and prevented any further damage, and they immediately admitted the error to maintenance.

The damage wasn't going to prevent the airplane to fly and was more cosmetic more than anything, but for the squadron, it was the principle of the problem.

Our commander's boss (the Group Commander) was not a C-130 aviator, and was one who didn't like to hear about damaged airplanes. Damage made to sleek fighters was much more problematic than for less-sleek and 'off-roading' airlifters. Other crews had also come back with some scratches from missions – sometimes because of stupid mistakes like ours. To the Group Commander, whose experiences flying never had to deal with a pebble on well paved runways, let alone completely austere LZs or abandoned FOBs or Afghan roads, a damaged airplane was heresy. Our squadron commander had to explain to him how C-130 operations were conducted in the rough and tumble of Afghanistan, and some damage was bound to happen: it was our job to go to these hard to reach places. We flew below 25,000 feet almost all the time. A fighter almost never flew below that. Our commander and his commander were both pilots, but of completely different worlds.

With that, though, our squadron commander still had to do his part to help diminish these mistakes. So he pulled us into the conference room and gave us a talk: don't let it happen again! He reminded us with a somber tone that no one notices our accomplishments: the airdrops, the humanitarian airlifts, etc. But they do notice the damaged airplanes and mistakes. The intent is to keep us safe – if we can minimize mistakes, we can minimize risks.

With that talking to, the team returned to flying and complete missions superbly. The commander had been very clear to say that he had absolutely no doubt in our abilities, nor was he suggesting that we stop executing the mission. He just wanted to convey the message passed along from his boss, and explain the differences in perspectives between the Group-level and our squadron. According to the squadron commander, we were still performing superbly, and he was more than proud of our abilities as a crew. John Hammond, the senior officer of the crew, was highly regarded and compared to some other crews, the squadron commander never had to worry about us – which was always a comfort.

But that's not to say we didn't face off with other challenges.

## A Night in the Airplane

After a 13 hour day, our crew was looking at our final stop before calling it a day.

Cruising at 24,000 feet, our crew looked at our In-Flight Guides and determined where in the sky we would start our descent to the airfield; the last FOB we would visit before returning to The Hub. We had a little bit of space open in the back for a few more boxes of material and we were scheduled to be on the ground for an hour.

The Sun was setting just as we began our descent into the FOB.

After touchdown, the plane was parked and the cargo was quickly delivered. Once loaded, and without skipping a beat, the loadmasters began preparations to start engines.

E: Clear number 1 engine.
L: Number 1 engine is clear.
P: Number 1 engine is…turning.
E: Light, drop, RPM, fuel flow, ignition…
L: Wow, wow, wow! Stop! Shut it down!

The pilot immediately let go of the ignition switch and aborted the engine start.

Over the roar of the other three engines, the loadmaster again described what he saw.

"There were flames bursting out of the exhaust. It was occasional, like sputtering."

The front-end crew looked at each other. Nothing on the gauges suggested anything was wrong with the engine. The other engines were running fine. Fuel flow was normal. No fire light was illuminated.

There was a possibility of what was called torching if it was extremely cold (-37 degrees C / -35 degrees F), but it was nowhere near that temperature at the FOB. The crew was flabbergasted. So we decided to try it again.

"NOPE! Shut it down!" again called out the loadmaster.

"Ok! Let's figure this out," said one of the pilots.

"It sounds like what he's talking about is torching," said someone else on the flight deck.

Pulling out the flight iPad, the crew searched for the problem and a hopeful solution in the aircraft manuals loaded onto the iPad.

Finding a reference in what the manual colloquially called the -1 (dash one), someone read aloud the instructions:

"During starting if the condition is reported before the ground start switch is released, place the engine condition lever to GND STOP and motor the engine with the starter. If flames spread beyond the tailpipe, follow the blah, blah, blah... Ok, so we can't go if we see torching, which it sounds awfully like this is what it is."

"You want me to take a look?" asked Carter. "I say that we give it one last shot before saying we're stuck."

Major Hammond thought about it and then nodded.

"Ok. Let's give it one more shot. Clear number 1 engine."

"Lieutenant Carter is clear. Number one is clear," confirmed Ford, the loadmaster.

Standing aside from the airplane, Ford stood in front of the propeller and Carter next to it.

Both again saw the flames bursting out of the engine's large exhaust pipe. The deep orange bursts of fire burst rapidly and weren't sustained, but loud bang were heard with them.

The loadmaster again over his radio signaled the termination of the engine shut-down.

We weren't going anywhere.

It was now time to determine what was next. The crew contacted headquarters in Qatar and our unit at The Hub. The plane would have to be repaired by a Maintenance Repair Team that would have to be airlifted to the base we were now stuck at. The airfield obviously didn't have the appropriate facilities or any qualified maintenance workers, so another C-130 would have to bring the gear and personnel to us.

Headquarters now had to see how this was going to happen. Every aircraft and mission already had missions slotted against them and everything was timed and coordinated properly to meet different restrictions. An Army operation was kicking off as soon as an airdrop was completed, or an airport closed at sunrise which meant gear had to be delivered before first light... And now they had a crew stranded at another base within hours of their 'drop-dead' time; the time at which we wouldn't be permitted to fly anymore for another 12 hours due to work-day and crew-rest restrictions. The race was on.

The race quickly ground to a halt.

Most of the missions already started were a long way from finishing their day. After all, we were supposed to be the next flight home, so it was going to be a few hours before another plane could be tasked. Headquarters made the correct determination that we were not the highest priority anymore. There was no way they could get a maintenance team ready in time to be loaded up, fly out and repair the airplane within 30 minutes, which is how long we now had before we would no longer be allowed to fly. It would take longer than that to get to us. Now, regardless of what happened, we were stuck at this base for at least the next 12 hours.

The crew started to seal up the airplane. While the officers gathered our classified materials and locked them in various compartments of the aircraft, the loadmasters locked the paratroop doors and closed the ramp and door. Once they were secured, we then locked up the crew entrance to the airplane and walked in the darkness to the building called Base Ops to find if there were any facilities for us to sleep in.

There weren't.

The small FOB had no additional quarters for the aircrew. That meant there was only one last place where we could sleep and get the crew rest required.

The airplane.

Our crew had a minimum of 12 hours before we would be allowed to fly again. For now, there was still no word about when someone would come repair our airplane so that we could leave this airfield. For all we knew, it could be 24 hours or more. Or less.

Before returning to our place of work to sleep, though, the crew walked around the FOB to keep busy for the next few hours before settling in for the night.

The base had a small tent with a mashup of seats, board games, old video game consoles, worn out pool tables and ultra-modern flat screen televisions. The televisions were being used to watch football games in loop. We sat in this morale-tent and did what we normally did back at The Hub. Small talk, laugh around before turning to our cellphones. We'd listen to music, watched saved movies and chat with friends and loved ones back in the United States. The morale-tent, of course, had Wi-Fi. Without Internet, there wouldn't be much morale for the 21st century warfighter, even at the most remote FOBs in Afghanistan.

With the Internet though, the mood still wasn't great among the crew. We were stranded at a foreign base, still in the warzone and held back by regulations telling us we had to wait and rest to fly. The crew all agreed that we could have flown home well after our 16 hours were up, especially since it would only be for one flight. The rules were the rules, though. Often reminded to us in flight school, those rules were 'written in blood.'

There was a feeling of abandonment. Headquarters didn't think we were a priority. Getting the other missions done was more important than getting an airplane fixed and off a FOB with its crew as soon as possible. We, of course, didn't know what was happening behind the scenes to get us home. All we knew was that we would have to sleep in our own plane, with no showers, no change of clothes, and others had a feeling that we somehow didn't get our job done. That we were distracting others from what they needed to do. Even then, they weren't coming to help us.

Our crew concocted various ways the schedule could have been changed for different aircraft and crews to get a maintenance team here. We all told ourselves that an aircraft could have just come to get the crew and left the airplane behind. We would come back with the maintenance team and a security forces detail to protect the aircraft during daylight hours. Instead, we had to stay with the plane, in the dark. This was still Afghanistan.

Those were the obvious frustrations of the crew, resulting from having to sleep in the airplane in a small place that didn't have facilities to spare.

The other challenge was more curious for the crew. In the end, the time off that resulted from being stuck is the same that would have existed back at The Hub. We would have also wasted time on our phones, flipping through old photographs, listening to music, watching movies or reading digital books. Not being home made us feel like we were wasting our time. Like what we were doing was just killing time, while doing these things The Hub felt more productive.

We were bored.

It was inexplicable. Perhaps it was because we weren't in our normal container-bedrooms to do the same activities. Maybe it was because we didn't have the same routine.

Back at The Hub, we all did our own thing after getting home. Maybe we'd go get dinner, or instead maybe we could take a shower. Where we were, this wasn't possible. It wasn't the first time nor the last time we would sleep aboard the aircraft – various opportunities and necessities presented themselves. This time though, it felt like being stuck in an airport after a flight was cancelled.

A very scary airport surrounded by barbed-wire and frequently bombed by mortar fire.

A few hours later, we quietly headed back to the airplane.

It was parked in the dark, close to the runway, surrounded by quietness. There was very little activity in the area. No lights, no sounds. It seemed that everything came to a standstill when darkness settled.

An advantage of being where we were was that there were very few lights, which meant we had a clear line of sight to a blanket of stars sparkling above us.

We opened the crew-entrance door to the C-130 and climbed into the plane. Inside, the crew began to put up some stanchion-seats; the canvas bench-seats used to carry paratroopers. McKay and Hammond decided to sleep on one of those rows each, and Telford would do the same at the rear of the aircraft. Sheppard decided he would break out his hammock and attach it to different points of the aircraft cargo compartment. Carter was going to take the bunk bed of the aircraft, while Ford would sleep in one of the pilot's seats.

The overhead escape hatch was opened to let air in, along with the pilot windows, and the crew settled in for the night.

Carter turned on the emergency escape light to read a little bit in the pitch darkness, while Ford used his phone to take some photographs. Sheppard also used his phone to read a little bit and McKay watched a little bit of a movie on his phone. All in silence, only the soft glow of their screens reflected on their faces.

Slowly, everyone fell asleep in the C-130.

The next morning, sunlight poured into the cockpit where Carter and Ford were sleeping. Ford had ended up wrapping his body around the throttle-body (where the four handles controlling the engine throttles were located). Carter had covered his face with his uniform blouse because of the mosquitoes buzzing around during the night.

As the day was slowly beginning with the Sun rising, we did our best to continue to sleep.

Sheppard grumbled as he shifted in his hammock. The loud buzzing that could suddenly be heard was frustrating. He was trying to sleep!

Thankfully, it soon disappeared and the crew continued to sleep as best they could.

A few minutes later, it was Carter that was awoken from his slumber by *incessant* banging.

Finally, Hammond opened the Crew Entrance Door to investigate, and saw the Maintenance Repair Team hard at work on the engine. They were attacking the prop with all the tools they had available, the beige and dusty world bathed in an orange hue as the Sun continued to rise.

After taking a few minutes to all wake-up, the crew looked at their watches and we saw that our 12 hours were almost up. The MRT was dropped off by another plane on their way to another destination just before we woke up. When the maintenance team was done repairing our propeller, we would be able to load them up and leave.

The front-enders asked how the weather at The Hub was, and upon hearing it was as sunny as the base we had spent the night at, decided to fly using visual flight rules back to The Hub. This would simplify life enormously, preventing any sort of challenges to try and connect to other agencies for flight plans to be filed. We still had the codes to permit us to fly in the combat airspace, so permissions weren't an issue. We'd load them into the aircraft systems and we would be free to fly.

That's exactly what we did.

After starting the troubled engine first, and without a problem, the other three engines were ignited and soon after, with everyone aboard, we took to the runway.

After take-off, we pointed the nose of the plane back to The Hub and flew home.

## Weather and Commander

The commander was flying with our crew. While this was an opportunity for the commander to come out and fly, be a part of the missions and see what the crews were exposed to on a daily basis, it also offered our commanding officer a chance to see how his crews were performing.

Another crew had unfortunately underwhelmed the leadership of our squadron, forcing the command staff to break up the crew in question, and piecemeal them into other crews, thereby breaking us up as well. Just like the flights over to Afghanistan, the squadron commander's intent was never to doubt the crew's performance nor hop on board choosing to evaluate a crewmember, but poor performance was hard to ignore by the ultimate leader of the unit. It was even harder to ignore the commander as the junior-most members of the unit that still had 'airmanship' skills to hone.

Our crew was the next crew to fly with the squadron commander.

While some aircrew members may have been nervous of flying with the squadron commander, it was often for good reason and without ever doubting the qualifications of their commanding officer. Lt Col Gaius, before being the squadron commander at the deployed expeditionary squadron, was the squadron commander at Clearwater and had served as the second in command (called a Director of Operations in the Air Force) at both Clearwater and the expeditionary squadron at The Hub even before that. Prior to being assigned as commander and director of operations, Gaius had been part of the Air Staff at the Pentagon following an assignment as a pilot for the prestigious $89^{th}$ Airlift Wing (the unit that flies the President of the United States using Air Force One). Before working at the $89^{th}$ and the Pentagon, he had served for years as a C-130 pilot, during which he performed the first of what would culminate to be 10 deployments to the Middle East. There certainly were no credibility concerns for our commanding officer when it came down to combat aviation.

Lt Col Gaius wasn't to board our mission for the whole day, only at the end of the day. Our crew took care of the majority of the combat airlift missions for the day and the commander was going join us for our last delivery. Headquarters had scheduled a stop at The Hub to offload drawdown troops and cargo before heading back out to Yala, and it was during this pit-stop that the Commander would climb aboard.

Yala was unfortunately not a runway designed to welcome the Space Shuttle, let alone anything bigger than a C-130. There was room for only one plane at a time at the small field bordering Pakistan, in a region that sees often battles between allied forces and enemy forces that took cover behind Pakistani lines. On our morning briefs, it was labelled as one of the most attack FOBs we could go to.

Fortunately, or naively, the crew wasn't more concerned than usual – the concern really was flying with the commander. This was a consequence of spending too much time in the presence of bombings and military risks in general: our perception of risk shifted toward irrationality.

Lt Col Gaius's boarding mid-mission, though, was an advantage for the younger members of the crew. The entire briefing and preflight portion of the day was no longer a concern and all the commander would do is hop on after we were already well into our day, settled into a good, rhythmic routine.

The commander got to our aircraft under the cover of darkness, his approach completely revealed by the powerful floodlights that lit the flight line of The Hub. The loadmasters were busy finishing the loading of cargo and soldiers being airlifted to Yala while the front-end crew began to start up the systems in anticipation of starting engines soon.

Climbing aboard, the squadron commander greeted everyone with a beaming smile but not any words or speech given how loud the power-cart outside was. Night had fallen and the commander was flying with us after a full day in the office. He settled into the left pilot seat and started to test his own oxygen systems and then plugged in his headset to join the internal communications network.

P: Test, test. How do you read me?
N: Got you loud and clear, how me?
P: Loud and clear. Alright! How's everyone doing?

Everyone replied affirmatively and we were soon ready to go.

P: Ok, since I just hopped onboard we'll do a quick re-brief of the upcoming mission. I'll be in the left seat, all other positions remain the same. Maj Hammond will still be up here with us and will be seated on the bunk… We're going out to Yala and back, any questions?

There were none so the crew got ready to go.

As we started engines, there was certain pride with the crew as we were able to show off how comfortable we were with this place that was the combat zone. We easily talked on the radios even though it sounded like an alien message being beamed to Earth. We heard on the radio that one of the taxiways was blocked, so we prepared the commander to expect a 'back-route' from Ground control. Ground told us to follow a 'back-route' and proved our awesomeness. We ran through the checklists smoothly, allowing this experienced pilot and officer to feel like he was among quality peers instead of younger, let-me-walk-at-your-pace flight crews.

Fifteen minutes later, we were airborne.

<u>SHANCS 55:</u> Departure, SHANCS 55, passing four thousand for eight, we'll be going tactical to the west.

<u>DEP:</u> SHANCS 55, radar contact, climb and maintain eight thousand, and is your destination Yala?

<u>SHANCS 55:</u> Affirm.

<u>DEP:</u> Understood, be advised, there's severe weather over the region.

<u>SHANCS 55:</u> Understood, then we'd like to request a holding pattern to the south of the capital area.

<u>DEP:</u> SHANCS 55, roger that. Climb and maintain flight level 1-8-0, fly heading 170, enter holding on the 187 radial at 16 DME to 26 DME.

<u>SHANCS 55:</u> Ok, flight level 180, heading 170 to hold on the 187 for 16 to 26 DME, SHANCS 55.

Some members of the crew sighed internally and the boost of confidence that existed on the ground vanished. This *had* to happen when the commander was flying with us.

Looking at the radar screens, Carter started to adjust the systems to get a better view of what was taking place in the Yala area.

We began orbiting about 65 miles from our destination, well within the range of the aircraft sensor systems. Looking over the documents handed to the crew at the start of the mission, there was no forecast for the weather we were now encountering. Which means we had no idea how long it was going to last.

Looking at the radar imagery, we could see two thunderstorm cells surrounding the airfield we were targeting. Ignoring the fact that we're not supposed to fly in thunderstorms, landing a heavy aircraft on a runway that was just drowned by thunderstorm-rainfall also created more challenges that needed to be considered.

McKay ran the numbers on landing on the small strip given our current weight, combining our fuel, cargo and aircraft weight. While he looked over his performance charts and calculators, Carter continued to study the storm using his systems and analyzed the fuel burn rate.

"So what do we do?" asked the commander.

The commander was waiting for a decision from the person usually considered the mission commander aboard the plane. It wasn't going to be the squadron commander, otherwise that would create a dangerous situation where everyone would agree with whatever he said (a concept in aviation crew resource management known as the halo-effect).

As far as Carter could tell, the storm was not moving. He switched back and forth between modes and adjusted the systems to start piercing through the clouds he was trying to detect. He wanted to instead see the ground and other features to take a positional fix. McKay determined that if we burned a few more thousand pounds of fuel, we could make it into the airfield and land safely, even if it was wet.

Carter was the counterbalance to that plan: if the storm was not clear by the time the weight was good, we would have to go back to The Hub. We could not wait longer, otherwise we'd run out of fuel on the way home. On the scale of bad to worse, running out of fuel mid-flight is worse than a storm.

So the aircraft continued to orbit while Carter tried to see if the storm could get clear during the 20 minutes it would take to fly to the airfield. The flight crew continued to run calculations and interpret screens that most wouldn't be able to make heads or tail of. Radar screen interpretation, or RSI, was an art that only the military was still trained in. The years of training were coming in handy.

"It looks like the weather is moving south," said Carter. The crew looked at the radar screens but we couldn't see it. Carter walked us through image that he was studying, showing various frames and demonstrated his logic and the crew nodded in agreement. "But it won't be out of the way in time for us to make the landing. Right now, it looks like we're going to be cancelling the mission and heading back to The Hub," explained Carter. "Give me a few more minutes and I'll let you know," finalized Carter before returning to the various computer systems and sensors, flight plans etc. He was drowning himself in his mission commander role to make sure the squadron commander aboard wouldn't doubt the decision.

Southern movement for the storm wasn't good, since there was more severe weather to the north. If it moved from north to south, then Yala would stay socked-in longer.

Carter studied the screens a little longer and the fuel burn while the commander continued to fly the plane, looking out the window at the darkness below, devoid of any light or signs of life – it looked like an infinite chasm of darkness. Sitting next to the commander didn't make the decision any easier. If it had just been the crew together, then the first lieutenant with a whopping 3 years of military experience would be comfortable making a decision. Our crew knew that we didn't skirt the mission rules or quit when the going got tough (the commander probably didn't doubt that either since we had taken more missions at the end of a busy day instead of quit for the night, and we maybe, kinda sorta came back from some missions with a damaged airplane once or twice). But the high ranking leader with 13 years of experience (a full decade more - or almost four times more) and the determining authority for the lieutenant's next assignment was sitting next to him. The LT better not mess up the decision.

Looking at the systems and at the master clock, crosschecking the fuel remaining, the flight plan and fuel burn rate, Carter confirmed his original decision and called it.

We were not going to Yala which could be considered a mission failure. The squadron commander acknowledged the decision and we got back on the radio with the controlling agencies to tell them that we were returning to The Hub.

The crew would find out within the next hour or so if it was the right decision, as the weather continued to move, or if it was a decision built out of an abundance of caution.

After landing and returning to ops to drop off the paperwork, checking on the military network, we could see the storm was still over the target airfield and it didn't look like it the system was going to pass anytime soon.

The crew's decision was vindicated.

And our crew was able to stick together. We didn't give any reasons for the squadron commander to doubt our abilities. Carter, for other actions throughout the month, earned an award for the month; it was not for his analysis and decision on that sortie, but for other accomplishments in-flight and on the ground.

After all, almost all pilots know not to fly into a thunderstorm.

## Happy Birthday

The Warning light was clear to the crew and McKay decided the risk was now too high.

E: That's it, we can't wait any longer. We *have* to shut it down.

P: Ok, let's do it. Confirm number 2?

E: Number 2, confirmed.

P: Number 2 engine, coming down.

The co-pilot pulled the lever and shut down the engine midflight.

MEDEVAC 55: Tetra, MEDEVAC 55, declaring an emergency, headed direct to 88XH3, switching to approach... Approach, MEDEVAC 55, declaring an emergency, 80 miles from Kandyland.

Approach: Copy, MEDEVAC 55, what are your intentions?

The crew relayed all the details over the radio to the airport that was our destination: Kandyland.

After landing without incident but with a watchful set of firetrucks lined up next to the runway.

Tower: MEDEVAC 55, if able, exit runway at taxiway of your choosing.

MEDEVAC 55: MEDEVAC 55, we'll take the next high-speed and shut-down there.

The crew maneuvered the giant propeller aircraft off the runway on an oblique taxiway and as soon as we crossed the two solid yellow lines preceding two dashed lines at the edge of the runway, we stopped the aircraft.

Firetrucks swarmed the aircraft and continued to look at the airplane as the crew exited and waited a little further away.

After the firefighters had finished their job and declared the area safe, the crew took their own time to look at the engine. Looking at the engine, among other things, the crew saw streaks of various fluids enveloping the cowling; those normally belonged inside the engine.

The bottom line was that we were out of the fight for now. A Maintenance Repair Team would have to be sent over, we were now stuck in Kandyland UFN (Until Further Notice), and it was one of our crewmembers birthday.

The wounded in the back were still able to receive care at this base, so part of our mission was accomplished – we had gotten injured people out of harm's way and back to a base that had adequate medical facilities. The aeromedical crew would find a way back to their primary base from this busy airfield, after they unloaded the evacuees from the back of our wounded bird. There were other C-130s that would stop by at some point, or a C-17 would certainly have spare room in the back for them (one of the advantages of being an aeromedical member was that you had a pretty high priority on the movement list).

Kandyland, fortunately, was a fairly large base and there were lodging facilities for transient personnel, which we now were. The base support services put us in similar containers that we had back at The Hub (but it wasn't the same – these containers didn't feel like our home containers, obviously).

Once we had picked our bunk beds and put our break-bags down (bags with overnight essentials that some of us learned to pack after our first unexpected night outside our base), we had the remainder of the day to discover the rest of the installation. There was the 'boardwalk' where dozens of small shops and restaurants flanked the outer edge of a wooden, square path that itself surrounded two small-sized soccer fields. These small shops and restaurants had offered a variety of things that we couldn't even get at The Hub, so the crew was certainly happy to be stuck here instead of a small FOB. There was a small European supermarket where Carter purchased boxes of Kinder chocolate and a bag of Chupa-Chups lollipops. Others got coffee from a coffee shop, and someone else got a smoothie from a smoothie shack. A few of us even got ourselves a celebratory cigar that we would smoke at the end of the evening, because even though we were grounded here, things had and were going well. There was a small, "shack-y" looking TGI Friday restaurant that served absolutely nothing like anything resembling TGI Friday "quality" food according to the people who lived in Kandyland. A well-known rest-stop even for the crew even on normal flying days where we weren't brought down for an unexpected stay was the Italian pizzeria. They sold pizzas that were of sufficiently high enough quality to satisfy everyone.

Later on and further down the other roads of the military base, we stumbled into the Base Exchange (BX), where their shelves were stacked similarly to those at The Hub, although this BX did have a small local man holding a cell-phone sale and repair shop.

Browsing through the shelves, dreaming of what we wanted instead of needed felt surprisingly normal. Maybe it was our consumerist impulses slowly returning to us, maybe it was a way to distract us from the abnormality of being at war which had become normal. The only reminder to us that we were at war when in one of the European supermarkets, clean, bright and stocked with products from home, were our guns and ammo magazines strapped to our legs.

Contrary to being stuck at another FOB, the crew was perfectly content and not concerned about being abandoned nor having to wait. There was no grumblings of being left behind, and perhaps this was because this base and airfield were so much larger and therefore more secure and robust. And even for dinner, we would be able to try something different than the typical food from our cafeteria at The Hub.

We went to the cafeteria belonging to the British armed forces, which was another unique experience. All around us, images of the queen, ranks with little crowns on them, and British accents. And Scottish accents. The food wasn't stellar, but it was different from our own usual meals at the cafeteria which meant it was still a great and exciting experience. It was crowded too, which meant it was lively. There were a few other Americans in the compound, but our crew still stood out.

When we exited the British DFAC, a soft glow continued to emanate on the horizon and around the lone rocky mountain point in the otherwise vast, dusty desert.

Being stranded in another airport was eerily similar to being stranded at The Hub. We quickly ran out of ideas for things to do. Other than hit the gym, there wasn't much else for us to do. The crew was on alert status, meaning that 12 hours after our landing, we could be called and expected to take off as soon as possible. At some point, we would have to get some sleep prior to that fateful call recalling us to operations.

So we returned to the *other* containers that served as our home for the night.

Before turning in though, a few of the crew gathered next to a small tree on the edge of the container stack that was the building we were lodged in.

There, they lit a cigar each and smoked a small birthday-cigar for the individual who was celebrating the day after a risky start. They had successfully evacuated battle wounded soldiers from the battle zone, even if they ended up forced to land at Kandyland, away from their 'home'. Those soldiers were going to survive though, and so did the aircrew, obviously. In the end, we had been able to visit a new base, too. A twisted way to be a tourist in a war-zone, but it certainly made things better after the hectic start to the day.

All in all, it had been a good birthday.

## Social Challenges

While deployed, most members of the military enjoyed a certain level of privileges that normally didn't apply to all other citizens or other non-deployed military. For example, we didn't have to worry about conducting our anti-chemical and nuclear warfare training, or mass-exercises... But there were other challenges that did appear and some aspects of life that we could not escape. Careers still needed to be mapped out, while harmony had to be maintained within the crew. All this while having to wonder why some actions were taken. Problems at home could bleed over to us (not just personal ones – big deals like the government).

There are three elements that usually affect moods between people: home-life difficulties, work-life difficulties, and interpersonal insensitivities (ie: I don't care about you).

Obviously being in Afghanistan didn't help the home-life. In some way or another, many of us had a difficult conversation with a loved one: "I love you, but imma go live in Afghanistan for several months. Good luck on your own." Some days did take their toll.

Work-life difficulties could be sourced from various situations: stressors from the actual job at the time, or concerns about future job security.

And when six guys were forced to spend months together, locked in either a cylindrical metal box (the plane) or a rectangular one (our container rooms), people could be sure everyone was going along great!

A common concern about future careers in the Air Force was amplified during our stay in the desert. While deployed, news of different units closing and other restructuring within the Air Force began to trickle down. First were the rumors – someone knew someone else who worked in the Air Force Personnel Center, where all decisions for careers were taken (if you wanted to be posted somewhere, or had desires to fly a certain airplane, it went through AFPC). Some rumors were farfetched, but others were not (although the Air Force's aims were still pretty farfetched). The Air Force wanted to cut thousands of jobs and would go through a Reduction in Force (RIF), which meant people's careers were more at risk then they had originally anticipated.

As official information started to become available, the commander of the unit was occasionally releasing information from the Air Force's Personnel

Center regarding military cuts that were going to be coming the following year. The breadth of criteria for elimination from the USAF meant that virtually everyone was at risk. Our unit leadership was powerless to reassure many, simply because they themselves were not told what was coming and how people would be selected for separation. The best commanders could do to help everyone was on a case by case basis, and overall to "keep-your-head-in-the-game."

We did, but there were still some concerns about what our future goals were. For those that were dedicated to staying in the military, they started to add-on more work to their already loaded schedules. They applied additional pressure to themselves to remain an attractive candidate to remain within the active Air Force.

These feelings and emotions also mixed into our larger inspirational drive. To become a military aviator, a certain amount of effort and motivation is also required. From our high school days, everything we did was evaluated and tested. Some went to the US Air Force Academy, others went to traditional universities, and others immediately enlisted in the Air Force. The result was still the same: to earn those wings, we had to meet the criteria set by the best and go through over a year of flight training and evaluations before we would be allowed to operate the C-130. To get there, you had to not suck.

A huge fear was not being the best. It was obvious that there would always be a majority of people who are *not* the best, but we were always striving to beat the odds and succeed. When contextualizing our positions, it can be evident why it is so hard to recognize how valuable and successful our efforts had already been: we were always surrounded by people who were in the same positions as we were and their accomplishments were the same. We were all highly trained. We were all conducting combat flights. Therefore, it was always a competition since we couldn't see anything else but the microcosm of the world we lived in: C-130 combat airlift. If we opened up a little the bubble, we'd realize we were already doing pretty well and accomplishing plenty. Some could try to blame our behavior on feelings of insecurity, or say we were not focusing on 'what's important in life'. But when we were in Afghanistan, reflecting on our cultural weaknesses and recognizing our societal pressures were not the first things we thought of. 'I wanna be the best, how do I make this happen,' usually topped the list.

Unfortunately, the spirit of competition was only increased now that the Air Force had announced a series of cuts. The games of politics were on.

If you were an aviator who was ready to separate from the USAF and you revealed such plans, then two questions asked themselves: would the commander really recognize your efforts, or simply give an award to someone who was expressing an interest in remaining within the ranks of the military to promote the loyal member?

People would start to focus on the missions being dropped down. Which would look better on a performance report? Which airdrop would help? Or a VIP flight could maybe push someone's career in a positive direction if they were able to talk to the VIP.

The young lieutenants didn't really know how to, nor really cared to, play these games. But when they saw their senior ranking officers starting to mumble about the assignments, they couldn't help but notice. When those office stressors were combined with the rumors of what others back in the United States were doing or accomplishing, some of the senior officer's desires to be in the combat area of operations started to dwindle.

This was normal and to be expected, as well. It was well known that the US military provided a very generous retirement program after only 20 years of military service. For many of these high-ranking officers that were not yet commanders, they were in a very vulnerable spot: they cost a lot to the military (higher salaries for higher ranks), were facing their next assignment being a 'staff job' (ie, they don't fly the airplane regularly anymore, so their skills atrophy and are no longer really utilized nor sought by the Air Force on a technical level), and they were exceedingly close to earning that retirement pension: another draining cost on the Department of Defense. From the perspective of these officers, they were the prime target in the USAF's crosshairs when it came down to making cost-effective cuts to the workforce. They had not hit their 18 year mark (where they are protected from forceful separations) and they had already dedicated over 12 or 13 years to the job with the intent to get to the 20 year retirement mark. The USAF could save a lot of money by cutting these higher ranking officers. The positive attitude of those higher ranking officers in Afghanistan was being put to the test and they now had to operate with that distraction in the back of their minds, with the other challenges also ever present.

The enlisted corps didn't play by the same rules, and their terms of service within the military were different than those of officers. While an officer's "commission is to continue in force during the pleasure of the President of the United States of America," the enlisted members who sign up with the military sign a "term of enlistment" which dictates the duration of their enlistment. They periodically apply to re-enlist if they want to continue to serve. Their career concerns were different, yet sourced from the same uncertainties. They would have to request permission to re-enlist in the Air Force, and if the USAF determined they weren't needed, then maybe they would be denied this re-enlistment. However, the enlisted were a little more discreet in their handling of the news compared to some of the higher ranking officers (Majors and such). Not only that, but most also had already terms of enlistment that took them past the timeline of the cuts, which meant that others would also be removed from service before their names would pop up.

Others were also seriously considering separation. Before removing involuntary members from the service, the Air Force was also coming out with a program known as the Voluntary Separation Program. This was not clear while we were deployed, but most suspected there was or would be a program such as this. Even less clear were the requirements to be eligible. Many pilots were willing to separate early and go work for the airlines, forever leaving behind the multi-month deployments. Officers had to sign a 10 year contract when they earned their wings, allowing the Air Force to recoup part of the multi-million dollar training provided to these combat aviator. This meant that officers have served 12 years in the military (when adding the 2 years of flight training) by the time that contract expires. If they had an option to cut that requirement short, many officers would have taken the option into consideration. But no word yet while we were in Afghanistan.

These difficulties often allowed crews and people to unite around (and against) a common narrative, however there were also those instances of competition that would cause divisions. Those and more.

Most already know the office-space interactions that develop can sometimes lead to tensions. A difference from what most Westerners experience though, was that we also lived with each other. Which meant that we were together almost all the time, every day, for many months on end.

Fortunately, we each did our jobs well which lowered the sources of tensions – we didn't bring home work issues back to our dorms.

Other crews were not so lucky. After one of the pilots executed a poor landing, one of the loadmasters mocked him. Regardless of whether or not it was done in good faith and just to be funny, the pilot didn't respond in such a funny way: "maybe if *you* were a pilot could you say something."

Even outside of our individual duties and responsibilities on the aircraft though, came the inevitable clashing of lifestyles. Some wanted to be social at all hours of the night and didn't think sleep was important. Instead, they stayed up all night before a flight to catch up online matters. Others were the polar opposite: sleep was sacred and light and noise were forbidden in our kingdom of rest.

So we decided to put those two together in a small container.

After resolving how to balance social and sleep came the obvious personality clashes. Some individuals were sometimes prone to the 'I have to be right and get the last word' mindset (especially when sleep deprived). Fortunately, none of us were ever sleep deprived and none of us had learned any lessons after being in relationships where this *never* happens. But it did happen.

We were also *never* put in high stress situations that tested our patience. But we were. Combined with our sleep deprivation, feelings of inadequacy and job cuts looming, there were days when some us were just a pleasure to have around.

The crew was still fairly good at containing the stressors and attitudes. Fairly. There was definitely room for improvement. Having those days off during which a roommate could vent without fear of having their true perspectives leaking out were an excellent way to relieve pressure. Trust was a vital factor for our team.

There were some unfortunate moments though where frustrations *between* the crew members did leak out. Carter could have sworn up and down that he had released the proper lock holding back a pallet, but it didn't matter. Telford still grumbled when he said Carter was incompetent to a point that the loadmaster had to walk to the front of the airplane and correct Carter. Normally, this was not a big deal; one mistake occurs. But with everything else piling on, even the most trained combat aviators sometimes let their frustrations show and a level of irrationality take control, sometimes causing some enlisted to toe the line of respect when addressing officers.

There had been a short period during the deployment where Ford had to be treated like a teenager because of his management (or lack therefore) of his sleep schedule. He would sometimes stay up all night to surf the web and connect with others in the United States through social media. Which meant he would get exhausted on a flight or miss the alert time altogether, forcing the crew to cover for him while he rushed to get ready. Hammond was forced to tell Telford to remind Ford to go bed – babysit him until he did.

Sleep deprived Ford would then become sensitive to statements that were less than delicate from others. Ford even got a crew-member from another crew involved involuntarily when he tried to prove Carter wrong. The public event showed insubordination bordering on childish from Ford toward Carter, and he unfortunately paid the price later on. But Carter and Ford still were able to mend the relationship after both recognizing that they were going through a period of particular stress for both of them (events from the war couple with realities of not being at home were taking a toll).

And while the team did have their moments of frustration with each other, everyone in the deployed team would sometimes suffer from the same inter-personal conflicts.

McKay was sometimes ridiculed by the maintainers for his stringent requirements. He would find himself left alone at the smoke pit sometimes, ostracized by others. It would take time for these behaviors to mend – it was part of the whole 'storming-norming' process. Even if it is normal to go through this process, it doesn't mean it didn't hurt at the time.

But we were still expected to be an effective team and accomplish the mission.

We did. Even with these occasional difficulties, the team was still able to exceed performance expectations. The missions were almost always accomplished and without any glaring problems. Maintainers still worked closely with McKay to make sure the planes could still fly. The commander didn't keep an eye on us, which was a good thing. Our abilities were not in doubt and our performance in Afghanistan was only helping our chances to stay within the Air Force. We were still very much able to keep the social troubles away from our professional responsibilities.

To the public eye, we were a good crew. Aeromedical teams flying with us admired our loadmasters and their methodology to help. Their work ethic impressed more than one, seeing that they never hesitated to roll up their sleeves and get dirty to get people and their stuff to wherever they were scheduled to go. The front end crew allowed visitors to see the cockpit whenever they wanted, and explained to them the principles of flight to anyone who wanted to listen.

Certainly, although the crew did go through a storming phase as just mentioned, it was only temporary and fleeting in the grand scheme of the time we were in Afghanistan. We still, for the large and vast majority of the time flying and living together, got along well and respectfully. There were some of us that just didn't mesh perfectly in the sense that we weren't going to become best friends, but we certainly were always an excellent team during missions that communicated very well and professionally. None of us ever found ourselves having to eat at the lunch table by ourselves, even on our off days.

That alone is something worth celebrating and acknowledging.

Our ability to understand that we didn't need to be inseparable friends as a crew made our crew surprisingly more cohesive than some others. In the moments our crew was united, for meals or in flight, we were still able to share laughs and tell stories that would make us all smile.

On a curious note, none of us ever publicly criticized another individual's abilities as an aircrew member. That we never took it to such a level was probably a saving grace compared to some other crews that sometimes tried to get people from other crews on their side (it only happened once from our crew and never again).

These temporary disagreements between the crewmembers were nothing compared to the rage we would sometimes feel against our headquarters in Qatar. These frustrations would occur a little bit more often than our interpersonal issues (way more often).

When some orders came down, our crew would unite like never before... against the CAOC (the Combined Air Operations Center, the headquarters in Qatar, pronounced kay-awk). What in the *world* were they thinking? Why the *hell* would we ever want to fly there if all we had to do was just fly to this place? The loadmasters would see some orders and say, 'today, we get to carry a tank that is thousands of pounds over the limits of our airplane, but other than that, we have a great day ahead of us.'

There were moments when Carter had to start talking back to headquarters and remind them of some basic common sense ideas. For example, 140 passengers would not fit on a C-130.

I, as the author, had the honor of working at the headquarters later on and was able to understand why some of those decisions had been taken. Sometimes they were for good reason, but there certainly were some moments when we dropped the ball when tracking the 50 or so daily missions we were coordinating from headquarters. Those were the times we needed the aircrews actually conducting the operations to call us and knock some sense into us.

But when we were flying the missions in Afghanistan, we were making NASA proud by once again proving their research correct.

Indeed, NASA has performed experiment after experiment regarding crew dynamics and the results of isolation. They found that space-crews could chose to release frustrations during long periods of isolated missions to one of three distinct entities: themselves, their crewmates, or mission control. Obviously, we were never responsible for our own problems (at least we told ourselves that) so we didn't vent nor blame ourselves for our frustrations. As our own experiences proved, there were sometimes when we growled at each other as crewmates, but fortunately, most of our annoyances were shouted at an inanimate object (such as a communications laptop) or through the radio (when the mic wasn't keyed), all directed toward our headquarters in Qatar (our equivalence to NASA's mission control).

We weren't as isolated as astronauts stuck in the International Space Station or the Apollo crews flying to the Moon, but we were constantly surrounded by the same individuals in a similar manner, and our usual group of friends and methods to avoid each other at home were unavailable to us in Afghanistan. We lived, flew and ate together. This officer the opportunities for great bonding, similar to the scenes you encounter in Hollywood dramas, but the reality was more subtle than that. We first had to pass through the 'storming' phase and learn to live with each other.

Which we fortunately did.

There are still plenty of positive moments we remember from our events out in Afghanistan (in case you hadn't noticed since you somehow continued reading to this page).

One day, as we started to pull out of our parking space at The Hub, Hammond listened to the loadmasters guide him back:

L: Keep coming back, turn towards 1... turn towards 1.... You can start your turn, so turn towards 1... please start your turn...

Hammond complied with the instruction and...

L: WOW! TURN TOWARDS 4, turn the other way, turn towards 4!

During another mission, Sheppard was getting impatient with how long it was taking for a team of army soldiers to get to the airplane.

SHANKS 55: Tower, this is SHANKS 55, I'm inquiring about the status of our upload.
Tower: Copy that, we'll contact ATOC and see what's up.

Immediately, the rest of the crew started saying that was a bad idea and that we didn't want to burn any bridges with the air terminal operations – if they were taking time, it was certainly for a good reason. So Sheppard immediately got on the radio again.

SHANKS 55: No... Don't!
Tower: Um... Well calling them is the only way I can find out for you.
SHANKS 55: Yeah, we're not going to get that fight start... Disregard.
Tower: So... What do you want us to do?
SHANKS 55: Do nothing. Sorry about that – there's training going on, on the flight deck.

Sheppard felt like he could have been more eloquent on the radio.

Just like all teams, our ability to mesh and integrate was put to the test. And just like all teams, we overcame those difficulties and successfully continued over several months the long missions that were the reason for our deployment.

## Combat Airdrop

Within our C-130 world, airdrops are very common. Most of our training flights include at least 2, sometimes 3 per sortie, and we fly about 3 sorties per week, which means we execute about 7 drops per week. We are all trained to perform a variety of airdrops, be it something as invaluable as parachutists or something like a cheap, low cost containers filled with water. No other aircraft in the Air Mobility world has 100% of their crews airdrop qualified. We take great pride in our ability to perform drops in any weather, on target and on time. As a parentheses, many of our Active Duty crews from our unit in Clearwater have also been invited by the Air National Guard to fly firefighting missions with them, having a direct impact on communities even at home, and also learning to fly some of the most difficult types of airdrops the aviation world has to offer.

So when we got the opportunity to perform a combat airdrop in a warzone, it was the pinnacle of our training and capabilities. It was the sum of everything we had trained for, since that moment we entered flight school. For some, it went even further back to when they first dreamed to fly. This was the complex combat resupply that sometimes made it into films.

We started as basic, undergraduate student pilots, learning to keep an airplane from crashing to the graduate level skills of hitting a target at night, flying at hundreds of miles per hour while accounting for winds that can blow the parachute toward our target point. All on time, every time. We did it almost every flight back at home, but it was rarely over dangerous areas. Combat added that slight danger.

The Tactics office was the brain of the operation. Traditionally, at home, the crew usually planned their own airdrops for their missions. They communicated directly with the person that was requesting the airdrop and worked out the details to ensure the mission was accomplished. But in Afghanistan, with crews always needed to fly, Tactics in the warzone was the planning office for everything, to give crews as much crew-rest as possible and maximize their utilization. The shop was usually manned by fellow crewmembers whose responsibilities were to plan most of our missions. The head of the office was the Tactics Chief, almost always a USAF Weapons Instructor School graduate; a grueling program that only the most dedicated flyers attended to earn a PhD level education and training about their specific aircraft and capabilities, and how they work in the big-picture of a war.

Consequently, they planned everything to the Drop Zones, down to every last detail. Whereas in the US, we went to trusted DZs that were certified and well surveyed, combat Drop Zones may have had more obstacles such as schools or hospitals, lakes or canyons that were in the area that needed the supplies. And then they had to plan for other unknowns that we, obviously, didn't know about. While the Tactics Chief was a senior ranking officer and a Weapons Instructor School graduate, the other two tacticians were any qualified officer flight crew-member. This small team of selected crew-members ended up flying fewer airdrops but they focused on making sure our plans bordered on perfect.

Late one night, our squadron ops received a phone call. A special operations task force called and said: we need this, here, and at this time. They sent a set of coordinates before the phone was hung up. The Tacticians immediately began to analyze the drop area using satellite images, and talking to troops that had been in the area before. With all the information they could absorb, the Tacticians came up with a plan to execute the airdrop as effectively as possible. The surrounding mountains were analyzed, the village, intelligence about the threats around the target were requested for headquarters, and relayed were the details of the mission we were supporting. Knowing what we were doing would help us assess alternate DZs, contingencies if the enemy decided to start shooting while we were on approach. All these details were fused into a plan. The crew chosen for the drop would be whichever one had their alert time closest to the target time requested. There was no need to consider various crews since we were all highly qualified to fly airdrops. So all crews had to do was wait for their turn. At some point, their schedule would line up with that of an airdrop.

The day prior to our first combat drop of the deployment, our crew was assembled in the Tactics office, where the Tactics Chief briefed in detail the mission. Again, most of the time, we simply did it the day of at home, but this still was a little different. We usually didn't have many communications issues at home, for example. If we couldn't talk to the drop zone in the combat zone, though, what should we do? Did that mean the troops are under attack and we're flying straight into a trap? Was it simply a bad radio? Or were we too low and we could be headed straight for a mountain peak in the middle of the night? The pre-day brief allowed the Tacticians to brief all the details without worrying about the timeline and causing the mission to be delayed if he took longer than planned to brief.

We went over everything, including a run-through of the mission with a simulated flight through our planned mission on a large flat screen in our wooden shack that we called Ops. The contrast was irresistible to point out. Working in a building similar and built with the same materials to those from the 9$^{th}$ century (a wooden shack in the desert), we observed computer generated imagery from spacecraft in orbit displayed on a flat screen hung on the wooden walls. The crew, lounging back in front of the flat screen with our checklists open, ran through the procedures verbally as if we were in the plane. This is what is colloquially called chair-flying in the flying community. The fly-through simulator played on the monitor without pausing, giving us the time-crunches we would experience in this unique environment. The crew did this to analyze and understand where some weaknesses may arise and prepare for them. We also went into other details, from radio frequencies, routes, to the altitude at which the doors would be opened, the rate at which the cargo would roll out, the contingencies if we had a problem with the drop sequence, if there was enemy fire, if there was another aircraft (most likely a helicopter) at our drop altitude... Two hours later, the crew was exhausted but we were finally done.

That one time brutal briefing was just that: one time. Future drops were less detailed in their briefing time and relied more on our previous training, with items simply being briefed as "same as last time."

The next day, drop day, started like any other day. Breakfast, AFE, Ops and our usual briefs. We looked at the schedule and it was the same as any other day except one of our destinations has the two letter identifier "DZ". But we were then rolling into a usual day of shuttling cargo and people between airports after the drop. It wasn't uncommon for us to be tasked with different types of mission over a day. The Tacticians requested that the enlisted exceptionally remain behind for the Tactics brief, doing a quick recap of the previous day's discussion. But all in all, it was the same.

We walked to the aircraft, again filled to its maximum capacity with containers, but this time, these had parachutes attached to them, and the entire load was rigged to fall out of the airplane in flight. Traditionally, something falling out of an airplane in flight was a recipe for disaster.

The drop tonight consisted of a mixture of fuel barrels and other supplies for a Special Forces contingent on the ground. One of the challenges (among many others) we were expecting was how to reach the unit over the radios. We were told we would have an additional communications-specialized aircraft in the sky to help relay information, but to still be prepared for potential issues. How well it would work, we would have to see. The engines were started and our day began like any other day.

Airborne, we switched to procedures that were more common in our daily missions at home in the United States, than in Afghanistan. A different set of checklists were opened, and the loadmasters were busy working the systems in the back (more than usual, at least). Up front, the flight deck crew tried to communicate with the drop zone. The first objective was to get to the zone on time: we were trained to a thirty second margin of error around the target time and we were going to meet that demand.

To make it on time, the crew adjusted speeds, cut corners on the air route if we were running a little late, or overshot a turn to lose some time if we were a little early. With a 40 minute flight, and unpredictably winds that could push us further back or forward in our timeline, there was a lot of air-math be performed on the flight deck.

Also activated was an air-corridor. This would allow us to fly without needing to coordinate priorities with Tetra. These air corridors were clear of other aircraft, giving the airdrop mission the priority it needed to get to the special ops team.

At a predetermined time from our target and our time-on-target, the crew was asked to begin preparations for the drop.

<u>N:</u> Crew, pre-slowdown checklist.
<u>L:</u> Acknowledged, Load.
<u>E:</u> Slowdown, drop zone, and escape?

N: No changes from the brief on the ground. We're a few seconds behind but we'll get there on time. McKay, we'll depressurize later since we're still over the mountains right now, but expect another 10 minutes before we can do that. If there are no questions, reviewed.

P: Pilot.

CP: Co.

E: Holding off on depressurization. CDS/ramp container bundle flap setting?

CP: Computed, 25.

E: Same. Altimeters?

P: Set, 29.92.

N: 29.92, SCNS set to 29.87 for the drop we'll reset the clocks in the descent, Nav.

E: Radar altimeter?

P: Set, 650, Pilot.

N: Set, 650, Nav.

E: Red light?

CP: On.

E: Computer airdrop information?

N: Checked.

E: AD/TJ (Airdrop/Troop Jump)?

CP: Set.

E: Pre-slowdown checks?

L: Complete, load.

E: Engineer, checklist complete.

As we got within 10 minutes to our time on target, the reality of the drop started to sink in. The jump lights were activated, and they glowed red. For now. We were all much more alert, double checking our systems and positioning, timing and cargo load. As was expected, communications with the DZ was challenging because of the mountainous terrain between us and the FOB, although we briefly were able to communicate with them through the use of the relay aircraft, orbiting high above us all. Regardless of the bad comms, we continued. We had coordinated different signals if we were unable to speak to the ground receivers and they knew them as well as we did. If the ground party sensed danger or needed to send a signal, we could do it covertly, without radios if need be, and without any outside observer noticing.

While we were more alert, the flight was not in any way more stressful or worrying. It in fact was comfortingly more familiar; similar to most missions we flew at home. We were just as quiet as we were throughout our other flights. A little busier, but not due to concerns: simply due to the added complexities of airdrops.

No one on the crew seemed particularly worried, or excited. It was just another mission that has been briefed in more detail than the others. And as we flew toward the target, the world was still acting like it did every day. Around us, we could still hear over the radios the other airplanes flying and performing their own operations. Fighters were being dispatched to a more violent region, and spy planes communicated with ground forces. Aerial tankers were set in their air-tracks, ready to help fighters stay in the air longer in order to support the ground troops. We heard some airplanes trying to fly in a direction that would take them through our own flightpath, but Tetra notified them that the airspace was hot; so the other planes adjusted their course around us. We continued to fly in our protected bubble toward our target. Our plane continued to rumble, and all the instruments glowed in the same green hue as they had before.

Unlike fighters or other strike platforms that like to finish things with a big 'boom', our objective was to be as discreet as we can be. The element of surprise and our ability to hide low to the ground were what made us useful. The team we were resupplying didn't want a large convoy of vehicle to be seen traveling on the roads, risking getting blown up. The vehicles would have also led someone following them to the team. Our plane could swoop in quietly and invisibly in the night, drop supplies in a few seconds and immediately sneak back out. No one would notice us and the ground team could continue its work in secret. Objectives, particularly those that involved capture orders or observational goals, sometimes took some time to accomplish and required the utmost discretion. A small team could hide its long-term goals by being observed with little supplies. An enemy would assume that the patrolling American forces weren't going to be out for long before returning to their base since they didn't looked well supplied. Little did they know that we could resupply them from the air, allowing them move wherever they wanted, without being encumbered by heavy gear and being able to call in tools as needed.

The drop was to take place at low altitude. Only a few hundred feet from the ground. Generally, it's not so much a problem, but the FOB was nestled within mountainous terrain; terrain that reached out of the ground for the skies. We navigated the plane lower and lower into the mountains, around peaks that started to rise above us. Without our advanced equipment, attempting to do this in the pitch black of night and in this hostile terrain and territory would border on suicidal. Fortunately, with our Night Vision Goggles, radars and other tools, we were able to navigate our way toward our target zone.

The goal was to fly low through the mountain and sneak out suddenly from the mountain gorges, drop the cargo and then shoot for the high altitudes of the sky, away from the target. Flying up the valley instead of through and around the mountain peaks was not recommended tactically: after all, that's where people lived, and we didn't want anyone to even catch a sniff of our presence. If those living in the valley heard us, they might call ahead and warn someone that we were coming. The rugged mountains didn't hold many villages or houses, especially in our flight path.

Two minutes out, we finally reached the edge of the mountains and stared at the valley, racing toward the drop zone.

Immediately, Carter called out to pull the throttles back completely.

<u>N:</u> Slowdown, slowdown now!
<u>P:</u> Flaps on speed.
<u>CP:</u> Flaps to 50% on speed.

After pulling the throttles completely back, the plane's warning horns came alive; the aircraft systems thought we were slowing down way too fast and we were going to fall out of the sky (in aviation, this is called stalling). But we knew what we were doing and the crew casually silenced the alarms as we ran the Slowdown Checklist. To prevent the parachutes from shredding on deployment open because of our high speed, we needed to slow down the airplane to a more manageable speed. This way, when the parachutes were deployed, they didn't rip apart because of the sudden rush of air.

Unfortunately, this normal part of the flight also made us more vulnerable, which is why we waited until the very last possible moment.

Carter had preplanned the slowdown point to be as close as possible to the drop zone, so that we would be slowed down just to the right speed just as we flew over the drop zone. That would minimize our time being vulnerable to small arms fire if there was an insurgent group nearby. Being slow meant we were more vulnerable. To be covert, we also didn't preemptively drop any flares – we would only drop the defensive flares if we saw something coming at us.

But nothing did as our airspeed bled off. Carter was satisfied: we had arrived at our slowdown point on time, which meant we would get to the drop point on time. The crew continued with the checklist.

> E: IFF/TCAS is set.
> CP: Flaps are set.
> E: AUX pump?
> CP: On.
> E: Ramp and door?
> L: Clear to open.
> E: Ramp and door indicates open. Flaps?
> CP: Set, 25.
> E: Slowdown checks?
> L: Complete, Load.
> E: Engineer. Checklist complete.

After having opened the ramp and door of the airplane, the loadmasters could see the ground rushing past us from the back of the plane. Within a minute, the checklists were almost complete. The plane was flying at hundreds of knots towards the target, which was about two miles out. In the dead of night, we could see through our Night Visions Goggles our surroundings. There was the FOB, a small hospital that we knew was to be avoided at all costs (that would be quite a surprise to open the hospital the next day and there's a hole in the roof left from boxes of cargo and parachutes), small *wadis* (dried river-beds that only flowed with water during the rainy season) which also had to be avoided, and a school. All the airplane lights were turned off to make sure that we couldn't be spotted by an observer on the ground, and in turn, shot at. In the dark, inky blackness of the night sky, the only clue that we were there was the soft hum of our propellers as we flew low over the ground toward the FOB. Our target time was also predicated on the time during which most of the population would be sleeping. Everything around us was immobile and unlit.

The plane continued toward the target: a small clearing surrounded by the previously mentioned buildings. The winds were analyzed again. Under the parachutes, the winds would push the cargo toward the target, so we had to make sure that we were positioned correctly so that given our altitude, the cargo would drift from the plane to hit the target. Through our NVGs, we saw a subtle signal from the SOF team that they were there and ready. Right on time.

Just like we were.

That was our signal that we were allowed to drop the supplies they needed. With ten seconds remaining, the final position adjustments were made – altitude, heading, airspeed and deck angle were all checked. And on time.

N: I've calculated a new CARP (Computed Air Release Point): 190 short, 55 left. Correct to the left now.

The plane was maneuvered with seconds to spare as the crew continued their sing-song, making sure the plane was on target.

N: Half a mile, 80 right correcting left. Good altitude. Good airspeed…. Five seconds!...

E: Deck angle is good.

N: Green light!

CP: On!

The loadmasters instantly saw the red light turn green and immediately sliced off the rope holding back the cargo. One by one, at an alarming speed, the bundles rolled out the back of the airplane, a cable attached to the airplane ripping their parachutes open as they fell away.

We felt the nose of the plane wanting to pitch up. The pilots tried to offset that tendency, keeping a deck angle so that the rest of the cargo could roll out, but without allowing the airplane to start climbing.

L: Load clear.
N: Red light! Right turn to a 0-4-0!
P: Flaps to 50, coming right, 0-4-0.
CP: Flaps coming to 50%.
E: Ramp and door?
L: Clear to close… Ramp and door is closed and locked.
E: Closed, door light out.
N: Start a climb and accelerate…now!

With the ramp and door closed, the plane could now fly much faster. Which was critical to get away from the area – we had been flying low, making some noise (however quiet the C-130 inherently was), and we didn't want our flight to give away the ground team's position. Not only that, but flying so low still made us vulnerable to surface-to-air fire. The higher we were, the more space we would have to maneuver in three dimensions to avoid the enemy missile if it was fired at us. We knew our threats, and the threats we were facing were designed to keep us from flying too low. Other threats exist to keep us from flying too high, but we didn't face those tonight. Tonight, we needed to fly faster and higher.

P: Flaps up.
CP: Copy, bringing flaps up.
E: IFF/TCAS is set. Radar altimeter?
P: Set, 200, Pilot.
N: Set, 200, Nav.
E: AD/TJ (Airdrop/Troop Jump) switch?
CP: Set.
E: Red light?
CP: On.
E: AUX pump?
CP: Off.
E: Completion of drop checks?
L: Complete, load.
E: And complete, Engineer… We're repressurized.

We climbed away quickly in the darkness of the night, quietly escaping the area, thinking back to what had just happened so quickly. The airplane was immediately reconfigured – the flaps moved, the doors closed and the plane accelerated away. In complete harmony, the crew was able to transform the airplane's shape and adjust its computers and engines while still flying at hundreds of miles an hour flanked by high mountains. We executed our usual checklists after the drop as casually as we have done throughout our training and other missions across the world. The loadmasters ensured the door could be shut, the plane was sealed up and we accelerated to escape the area and climb away from the ground.

The time between the green light turned on and then turned red again was 4.3 seconds. Years of training and selection boards, starting all the way back to high school, exhausting flight examinations, calculations and exercises, tireless discussions and long hours of airdrop theory and experiences came down to those few seconds. It was all done to ensure that those 4.3 seconds took place on time, and on target. The list of things that could have gone wrong could fill encyclopedias; this is why it's important to note that everyone involved in the operation was critical – if someone had not been there to properly rig the parachutes, or a weatherman had given the wrong forecast, the situation could have ended very differently.

The special operations team was resupplied properly and we could, so far, declare mission accomplished.

As we rushed away from the ground, we typed a message with our onboard computer to headquarters in Qatar to let them know that it all went well, and we were headed back to The Hub. Drop complete. Like the way over, there was no exaltation. No cheers of adulation to each other. No high-fives or chest-bumps. The mission was very normal and professional. We all knew we performed the job we were asked to do, and we completed it well. That was it. If it had been a monkey in the pilot seat with a moose as a loadmaster, then that would deserve applause. But the government would be happy to know that the financial investment placed in us over the years was not for naught tonight (the average cost to train an Air Force officer to fly is about $3 million, not including officer training or university costs.)

The return to The Hub was as uneventful as all other returns – terrible radio communications made the flight still as exciting as the low level drop. But we were soon back on the ground, taxiing back to our parking spot. Just like all other missions, the maintenance crews rushed to the plane to remove the little airdrop equipment we had and to prepare the airplane to accept passengers and cargo.

The officers went to the Ops building to ensure that there weren't any updates to the night's plan. We still had three more sorties today. Troops had to be transported from The Hub to another FOB, and then an aero-medical mission would be flown. Within thirty minutes, the crew stepped back to the airplane to continue our night like any other.

While this is the recollection of a planned drop, crewmembers from this team would later find themselves performing on-call airdrops, where the aircraft received a set of coordinates for the drop while already airborne, and then planned for the drop while enroute to the DZ. This was often to support troops in contact (TIC) or for other 'immediate' taskers that were time sensitive caused usually by enemy activity.

Members of the crew would also be conducting humanitarian airdrops later on, dropping thousands of MREs (Meals Ready to Eat) to refugees surrounded by flood waters or by enemies. It was our training and the strength in character of the crew that allowed us to be so versatile in order to conduct operations that are still not completely open to the public.

## Peaceful Nights

Quite often, the most peaceful times during out deployment would not be the dorm rooms, but the airplane itself. Flying from The Hub to Klein Deutschland is an example of those times, particularly as we flew in the middle of the night, when much of the chaos in the air (and on the ground) would be paused until the next morning. This northerly destination was also famous for its cafeteria, which made the travel to that destination even more attractive.

On one of those nights, we were flying back to The Hub after several missions around Afghanistan, culminating to our 12$^{th}$ hour on duty. At 23h30, the entire country was enveloped in the darkness of night. The airplane was fairly quiet, and so were the radios. Occasionally we'd hear a drone operating or we'd cross paths with a C-17 cargo airplane, but otherwise, everyone it seemed, was fast asleep across much of the country.

Like most of our missions, we landed early at The Hub from our previous destination, and were looking to take-off early as well to head to our final destination, Klein Deutschland, before being done for the night. Normally, the crew would swap in and out of the plane to go grab some food at the DFAC, but our next destination made taking off earlier a priority over getting food at The Hub. This was because the cafeteria at Klein Deutschland was famous for its delicious, European meals. Consequently, while the Loads would offload and onload as quickly as they could, McKay focused on getting the airplane fueled while the officers headed inside to Ops to make sure everything was ready for our arrival at Klein Deutschland.

Turn time (the time it took to land at The Hub and then 'turn around' to take off again) ended up being only 45 minutes, which was pretty good. Carter coordinated and got Positive Launch approval from Qatar, and the crew started engines. Backing out of the parking spot and taxiing to the runway to take-off, The Hub was only slightly more active than the rest of Afghanistan even in the dead of night. Heading toward the runway, our crew ran the same checklists as we had for past 5 sorties of the day.

The plane pulled onto the runway and the throttles were pushed to their take-off power setting. Rolling down the long runway, we soon found ourselves airborne and climbing away from the ground.

Airborne, the airplane turned westbound through a long valley as we climbed up from our paltry 3,000 feet to 17,000 feet. Although our destination was north, the climb westward was necessary because of the long and tall mountain range rising up to 16,000 feet that stood between the airplane and Klein Deutschland. We still had to get to our 17,000 foot level that would allow us to 'hop' over the sea of mountains.

Sheppard signed off with departure immediately after take-off because our westward path was nowhere near the busier airspaces; we didn't need to be controlled by an agency. We were flying into the most mountainous places Afghanistan had to offer, and there was no need to be guided because there were no viable threats in the neighborhood. Carter quickly got in touch with Command and Control, and after telling them where they were headed, Tetra had nothing to tell the airplane. Five minutes into flight, and the trip was already shaping up to be peaceful – we didn't want to talk to anyone and no one wanted to talk to us.

It was once we leveled off that things really sank in. Everything was quiet, dark and serene. This was a stark contrast to even ten minutes earlier when we were still talking on radios, but more significantly, hearing everyone else on the radios as well. Below us, an unending sea of snow-covered mountain peaks, and nothing else. There were no trees, not even small lakes, let alone signs of civilizations. There weren't even radio relays which meant we no longer heard anyone; the mountains blocked off most of the terrestrial signals. We could connect to our satellite communications (SATCOM) systems, but it wasn't necessary for our northerly destination, so to save bandwidth for actual users, we didn't do so. Looking out onto the landscape as we slowly flew over it, the Moon brightening the bright-white snowy peaks and amplifying the darkness of the chasms between said peaks, was an inspiring moment. Moving created the illusion that the sharp peaks and razor edged crests slowly waved, like foamy wave caps.

It was very different from the scenery above us. The cloudless nights allowed you to see almost as many stars as the sky could hold. While not as impressive as the view crossing the ocean, the black sky still scintillated more than any clear sky in any other country we had been in. Using NVGs, the sight was just as impressive, but this time in a shade of dark green, and the stars lighting up our eyes brightly.

As we travelled to Klein Deutschland, thought, we hardly used our NVGs, instead just keeping them by us at the ready. At the end of the long day, everyone was quiet and the silent, majestic yet soulless landscape around us only emphasized more our willingness to stay quiet. The loads lay down on pallets or stanchions seats as we made the crossing over the rugged void. The front end crew members remained in our seats, the soft glow of instruments being the only source of light on our faces. Some of us had music playing into one ear with a headphone plugged into an MP3. The autopilot was flying, and all the crew had to do was monitor the airplanes' capability to fly itself, which it did flawlessly. For the crew, this was our time to rest and take a breath from the day, even though we weren't done yet. But for once, thanks to where we were in Afghanistan, we didn't need to worry about flying tactically, or conjecturing up a surprise approach to a tiny dirt landing zone. All we had to do was relax in silence and look at the stillness around us.

For forty minutes, the crew stayed quiet for the most part. We had been chatting and kept excited for the majority of our 12 hour day. We had been running around The Hub to get info, or helping load cargo at the previous bases beforehand. Now, as we transitioned over the dark, bottomless canyons were flanked by the bright snowcaps. On the horizon, the triangular, jaggedly shaped snow looked like it hovered in the blackness of the night.

It was an irony that as we were travelling at 350 miles per hour over the natural barriers that are mountains, we were lost in thoughts in one of the most stagnant and still moments of our day. We looked down on the vast darkness below and the darkness above and remembered that we were a simple, dim dot of light flying on a blue sphere orbiting a star as insignificant as the countless others we saw above us.

This peaceful journey lasted about 40 minutes, before we started to come within range of the radio broadcasts from Klein Deutschland. The base was nestled right behind, 3 miles, from an 8,000 foot mountain chain, the last of the peaks that began at The Hub, 40 minutes behind us. What this meant was that we had to get close to the base before we got any sort of information regarding winds, runways in use and other information regarding threats or our mission.

Those mountains were right in between the plane and the base, and radio signals to and from these two couldn't penetrate the mountain.

McKay was listening intently on the radios for the automated broadcasts the installation had, making out numbers within the static caused by the mountain range, calculating the aircraft's capabilities for landing. The rest of the crew was preparing to land, building programs into the flight computers and calculating their descent rate and profiles to make sure we could land on their first attempt.

A few minutes later, controllers could finally be heard and communicated with, ending the peaceful crossing and bringing our crew back to the realities that we had to land the plane. With that, we ran the checklists and other pre-landing procedures.

E: TOLD (Take-off and landing data)?

CP: Checked.

E: Crew briefing?

P: Left seat landing, speeds will be 131, 114, 100. If I'm not down by the 4 board, send me around and punch me for my utter failure. Weather's fine. Complete.

E: SCNS (skins) LZ data?

N: You've got it loaded on a 3 degree glideslope, if you want it.

P: Thanks.

E: Altimeters?

P: Set, 2992, Pilot.

N: Set, 2992, the field is calling 2994, SCNS set, Nav.

P: Agreed. Thanks.

E: Radar altimeters?

P: Set, 200, Pilot.

N: Set 200, Nav.

E: GPWS (jip-wizz)?

CP: Set, Co.

E: Vertical Reference switches?

P: Set, Pilot.

CP: Set, Co.

E: Defensive systems.

N: Currently active and set. Will change systems when we get below the usual altitude.

E: Descent checks?

CP: Complete, Co.

N: Nav.

L: Load.

E: Engineer. Checklist complete.

Like everywhere else in Afghanistan, the air base we were about to land at was also very quiet at this hour, and we were the only ones coming in. Talking on the radio took only a few minutes: we called and said hi, requested to land, and we were given permission to land, with very little restrictions on our flight path to get to the runway. That was it: the crew was again only talking to each other, discussing our profile and preparing to descend.

Around 15 miles out from our landing point, or about 5 or so minutes before touching down, we began our descent from our cruise altitude of 20 something thousand feet. The pilot-flying pulled back the throttles to flight idle (which means as far back as possible without putting the engines into ground idle or reverse) and aimed the plane's nose toward the ground. The plane dove toward the Earth. Klein Deutschland was a well-lit airport that we could see from fairly far out but we still had the 8,000 foot range that had been an issue earlier for communications. Our dive was planned to have us cross over the peaks just 1,000 feet under us and of clearance. After crossing over the jagged obstacle, we continued the dive on the far side of the mountain and maneuvered to line up with the runway. The plunge was impressive, and with no one speaking on the radios nor anyone speaking on the airplane except for the occasional update on our performance, the loudest reminder of the maneuver into darkness was the loud wind-rush surrounding the airplane, resulting from our high speed nose-dive.

Once we descended to our pattern altitude of 1,000 feet above the ground, the crew finally pulled up to level off, slowing down enough to drop the landing gear and make sure everything was set for the landing. All of these events took place in the utmost quiet possible. Unless there was a checklist item to respond to or a question about our flight profile, no one spoke. The radios were quiet, and apart from the lit airfield, the rest of the world was dark and silent.

The plane touched down uneventfully and we taxied off the runway where controllers released us to the Follow-Me truck. The Mercedes van, a change from the usual Ford pick-up trucks seen at other bases, guided us to a parking spot on the NATO ISAF ramp. As usual, we parked right next to a NATO AWACS aircraft that seemingly never flew (at least it was always there when we showed up, and we showed up as much as we could for food).

After arriving at our typical parking spot on the ramp and shutting down the engines, the doors opened and the only sounds emanating were from the back of the airplane where the loadmasters were guiding in the forklift. Carter was at the front to control the pallet locking control mechanism and McKay took a walk around the airplane and make sure everything looked ok before going to smoke on the grass next to the taxiways.

Once the cargo was unloaded and new pallets were loaded, the best part of the evening began.

Some of the crew left the quiet flight line to go grab some dinner (even though it was 01h30 in the morning) since we were early from the planned timeline and headquarters in Qatar did not expect us to take off for another hour or so. Two people stayed at the airplane while the other four walked out. The walk from the airplane to the Entry Control Point was blissfully quiet – the ATOC operators were long gone with their cargo, and the bright spotlights that illuminated the parking ramp made it clear we were the only people around.

And the same held true as we walked to the Dining Facility that was two blocks away from the ECP; the entry point to the flight line area. After saying 'guten-morgen' to the German soldiers manning the ECP, we walked along the sidewalk, the only other person we passed was a German soldier on his bicycle - another rare sight in Afghanistan. However it was sights like these that showed the diversity of the world we lived in. We were American Airmen walking in a German base where European vehicles were driven and some even rode their bikes as they were used to in Europe, all while in Afghanistan.

The cafeteria was open from 1 AM to 10 AM for breakfast, which meant it was open when we got there. The particularity of this cafeteria was that it was the only one in Afghanistan to which the flight crew had access to that allowed diners to eat in actual ceramic plates, with real silverware and drink using glass cups. There were no plastic utensils, nor cans to drink out of. Just fountain machines, and delicious food that was hard to find anywhere else in Afghanistan.

Everyone helped themselves to bread, baguette like rolls, European butter, salami, and more. We could drink hot-chocolates, cappuccinos and other sorts of coffees from the machine located next to the soda fountain machine, which served peach-flavored ice tea, and other flavors unheard of in the rest of the deployed environment.

Contrary to most cafeterias in the country, this cafeteria was frequented more by European members than anywhere else (except maybe Gobble), and it was staffed by European soldiers and contractors. The cafeteria was always very clean and quiet. No one spoke loudly. Combined with the use of silverware and china plates, this eating experience was always unique. It was during these rare moments that we could sit at a table quietly and eat with what most Americans would consider normal dishware. These small moments were islands of memories where we didn't feel like we were necessarily in a deployed environment – where a sense of normality would be reinjected into our memories. The quiet flight over, and now a quiet meal would remind us of the simple things that were normal at home and that we had forgotten about over long months in a new world. There were no loud TVs blaring the latest football games from AFN/ESPN, Marines laughing loudly or Army soldiers shouting their latest prank on a fellow soldier. The loudest people here were the American Air Force Airmen that had arrived with their C-130 (that would be us).

The only downside to this dining facility was that we couldn't take food to go, since everything was served on trays and real plates. We asked for, and then rigged plastic plates handed to us by some of the staff into carrying shells in order to get food for the crew members that stayed at the airplane (the staff thought this was very strange). Most times, we would rotate the roles of who stayed and who went to this glorious cafeteria, which was felt as luxurious as a

French Michelin starred restaurant after our long months of the dining facilities at The Hub and elsewhere.

As we wandered back to the airplane after eating, we saw all the different details of this base compared to the others. European armored vehicles, more Volkswagens and Mercedes instead of Ford or Dodge trucks.

Motivation to leave for The Hub was hard to come by, which meant we instead relaxed at the back of the plane while the two crewmembers who had stayed behind ate. We requested Positive Launch with headquarters in Qatar, and while we waited for their response, we continued to tell each other stories, laugh and share with each other these moments that we couldn't share with anyone else.

But the word did eventually come down and we were cleared to launch and head home for the day.

Most of the pre-flight checks were still complete since a power-cart supplied the aircraft with power throughout our time on the ground. There was no need to restart and reprogram the electrical systems since the cart provided the necessary power, so we just rolled right into the Before Starting Engines checklist and quickly got to starting the engines. Ten minutes later, we were lined up on the runway, ready to take off.

After taking off, the trip home would continue to diminish our focus and work mindset. We climbed parallel to the 8,000 foot ridgeline and then hopped over it, heading back south toward The Hub. Our flight path would parallel an air-route that civilian jets used, but none were airborne at this time. We had just eaten, and were headed back across the hundreds of miles of mountain peaks. The void across was just was quiet as it had been coming over to Klein Deutschland. The night was still pitch dark, and the stars were out, the Moon having set though, making them even brighter and more visible.

The silent transit was just was peaceful. Nothing but the vibrations of the engines could be heard and the instrument dials were the only source of light within the cabin. Going to the back to use the restroom continued to emphasize the end of this long day, the loads sleeping on the pallets, the windows to the outside world showing nothing but a faint outline of the sharply rugged terrain below us.

It was during these moments that we reflected on what we were doing – what our place was in this war. None of us on the crew had any illusions about what the reality of our role in this conflict was. We were trash haulers, for all intents and purposes. None of us on the crew tried to call it anything else (although some in our squadron back in Clearwater sometimes did embellish a little bit). Hammond's humility spread to the rest of the crew. This allowed our team to be respected as we interacted with different users (what we called the people or organizations that hired us to carry them or their cargo into different LZs). We delivered the cargo.

Fortunately, one man's trash is another man's treasure. We certainly had no purpose for MRAP parts or grenades while flying up at 24,000 feet, but there were others who did need it. The crew and our role as mobility airlifters helped open our eyes to the reality of the logistic involved to ensure people can do the work they must. Granted, we generally didn't actually drop the bombs that make for great establishing shots in the news or film action scenes, with the occasional airdrop exceptions. But between aeromedical evacuations and the overall supply of materials across the world, we did ensure those action shots could be performed.

The pride most of the crew found was that they were able to quietly work with an absolute confidence in our ability to do almost anything requested (as long as it was structurally possible for the plane).

Our training was what made us so effective. Being effective was what the military needed, not necessarily efficient. If a team on the ground needed a single pallet of munitions and water, we weren't going to wait for the rest of the plane to fill up before going to deliver it.

So while we flew in our propeller airplane over the dark mountains of Afghanistan, we knew full well that were driving a cargo truck. But what we could do with that truck was what was unique and required multiple layers of testing and rigorous screening by the military training programs. While most of what we did was benign, we could also deliver a package to a doorstep.

Our skills allowed the nation and our allies to conduct power projection. The motto on our sleeve back at home was "anytime Anywhere."

Other times we would often wonder about the future. What was next in life? After flying combat airlift aircraft into various LZs and conducting airdrops while working on special operations, seeing the world... Some of us, like Hammond, were considering making the military a long term career. Serve the 'full 20' as it was called, since it took 20 years of service before one could retire with a pension and full benefits. Others wanted one thing and that was to leave the military – start a family and a normal life. Others were undecided.

Toward the end of the flight, we got an idea of what others had been thinking about during the flight. It was made evident through the questions asked or the statements.

"At the end of this mission, I'll have earned the leather jacket," said Ford through the headset, over the intercom.

He was right. Per the traditions set long ago, an enlisted aviator had to fly a certain amount of combat missions to earn the 'right' to wear the traditional leather bomber-jacket many aviators wore. This was another proud achievement for him.

"Are you going to get your blood-chit sown on the back?" asked Hammond.

A blood-chit was a tool all aircrew members had on them in the event they find themselves stranded in enemy territory. A single sheet of paper made of Tyvec, had a large American flag printed on it with underneath, written in English and several other languages (usually the one for the battlezone we were in), a message indicating that the U.S. will reward anyone assisting the bearer to safety. The blood-chit represented a written promise – an obligation of the US Government, which is why it was valued by the aircrew member: they could entice safe-passage back to allied lines to a willing local. The US Government *would* reward the person who helped – that guarantee was what made the sheet of paper so powerful. Rewards could vary; our western capitalistic minds immediately think of money but some wanted herd animals, or water filters, or medicine for a sick family member that saw their medicinal stores hoarded by a warlord.

"I don't know..." admitted Ford.

He returned to his thoughts, and the intercom between the crew once again went quiet. Some returned to listening to the music from a headphone slipped underneath one cup of their headset (or through the Bluetooth system if the crew-member had a fancier headset). Others just continued to stare out into the infinite universe, reflecting. Some definitely felt the end of the day coming – a thin veil of fatigue settling in.

Some people sometimes found long flights tiring. It was an ironic discrepancy that has always been a mystery for most of us, crewmembers. We could actively fly the airplane for hours, staying busy, but the moment we sat back and turn on the auto-pilot is when fatigue (and boredom) started to surface.

It should be noted though, that flying the C-130 was not like flying in a fancy jet airliner. Some crewmembers suspected the vibrations from the engines made them more tired than they should be. None one had any suspicions about the amount of noise generated by the propellers known as the "four-fans-of-freedom." That level noise meant we wore headsets or helmets throughout our 14 or 16 hour flights, along with ear-foamies below that as well, to block out as much noise as possible. The constant pressurization and depressurization of the aircraft as we descended to landing zones or for airdrops, and climbed back up 20 minutes later, was also tiring.

To be more precise: it was exhilarating, is what it was. Those things were never thought of unless we got bored on long, long flights. And even then, we were usually thankful to be able to do a little bit of nothing for once. We could finally focus on catching up on our flight logs, take a look at the view (instead of our knee boards to study the next approach), or write notes that would end up in a memoir. Ugh!

If we were over the United States, flying our New York to San Francisco shuttle for the sixth time in a row, maybe things would be different. But here, isolated for just a few moments from the commotion of the warzone, these moments of silence were golden refuges of peace, and no one ever complained of they being tiring. Quite the contrary, they were reinvigorating. They helped us recage and refresh for the 'return' to The Hub and all of its tumultuous activities.

Another 40 minutes later, we started to hear the radios crackle to life as different bits could break through the mountain ridgelines. The airspace was much busier than Klein Deutschland, and we had to be more alert this time because we would not be allowed to fly however we wanted to land. As mentioned before, The Hub was located in a bowl surrounded by 15 to 17,000 foot peaks, which meant we would be once again diving to land, similar to Klein Deutschland. Similar, as well, was the fact that The Hub was well lit. Always well lit – you could see from any altitude the perfect outline of the base as floodlights surrounded the entire compound, looking like a ring of light from above.

As we started to communicate with the different agencies, approach, Tetra and Command Post, the flight deck became alive again. While we began our high speed descent, we were searching for different aircraft around us: helicopters were moving below us, drones were operating just next to the runway at the same altitude as us, and a few other planes were also looking to land. We were getting in line to wait for our turn. The bustling airspace was a reminder that we were back in the more 'familiar' and active Afghanistan. Everywhere around us, an enemy could take a shot at us, and there were dozens of airplanes around us that we could collide with. The lazy-mindset that had afflicted us was reluctantly being shed with every instruction given to us by the controllers, reminding us that we had to be precise in our movements. The quiet night was over and we were back to a normal mission, regardless of the time of the night.

Fifteen minutes later, we landed without any problems and our routine returned. As we talked to ground, we were reminded how much busier The Hub was compared to Klein Deutschland; a fact often forgotten regardless of how often we flew back to that peaceful realm that was Klein Deutschland. The activity at The Hub, with cars and trucks driving around the parking spots and other airplanes taking off and landing was still quiet, relative to other portions of the day, but much more active than an hour before, up north. We were no longer the only plane moving. A helicopter zoomed above us as it headed out for a mission while we waited for a small spy plane to cross an intersection. Their days were just beginning. Thankfully for us, our day was virtually over.

We parked and shut down, and our routine to end the day resumed. As we packed all of our gear, our 14th hour of the workday was ending. As the loads unloaded the cargo, the front-enders shut-down systems and filled out the paperwork of the flights and goals accomplished.

Even after 7 legs of flight, and months of deployment, McKay and Carter still diligently compared their notes for the flight. Take-off times, cargo carried, fuel loads etc. They ran through their numbers and calculations in detail. If there were discrepancies between their results, they would often talk and figure out where the source of the inconsistency was. They acted like a team. Even though they all wanted to go home, the team still did their job without being told – it showed the remarkable discipline of these military members even after a lazy last few flight hours.

The maintenance crews were cool and told us that the airplane didn't need to be re-rigged for different cargo since the next crew was going to head out to a similar mission. That was one less thing for us to do. The trash bags were sealed and the back was swept with a broom and fifteen minutes after shutting down engines, the cargo ramp and door was sealed permanently for us.

All six of us walked back together to the AFE cabin where we dropped off our NVGs and other flight gear before heading back to Ops where different crews were preparing to depart on their first missions of their day. Our day had ended, but the war continued to rage, except over the 200 miles of 17,000 foot high lifeless, rocky, jagged, peaks that tapered down to 8,000 feet, around a more peaceful world.

Other peaceful runs were the long flights to and from Little Italy or Dandyland and The Hub. Even longer, peaceful flights were missions to other nations. To fly to the Persian Gulf nations, we would settle in on the PIFR (pronounced 'pif-er' – Pre-coordinated Instrument Flight Route) which was an air corridor through Pakistan that didn't require DIPS (Diplomatic Clearances). On this air-highway, we would hear the various bombers and tankers flying high above us, all of us knowing the Pakistani government was studying our flight paths very closely. If we deviated outside the PIFR corridor (or the 'Boulevard' if we were flying without an IFR clearance), the Pakistanis would notify the American Embassy and charge a hefty fine to the US Government.

These long flights were a change from the usual hectic profile of our other flights: calculations of flight performance, tactical approaches, changes of configuration in the back of the plane, pushing pallets in or out of the plane every 20 minutes, the constant radio chatter, the attention needed to prevent a collision...

Instead, the long flights were a moment of peace. A moment for the entire crew to laugh together, ask each other questions and bond, or simply rest, and quietly let our imaginations travel away from Afghanistan and the war we were supporting.

## The Closing

Everything that has a beginning must have an end. Most people would disagree to that point when talking about the war in Afghanistan, but at the time our crew was out in Afghanistan, the President had started to order the retrograde of US and NATO forces from the Asian country. Part of this retrograde was a diminishing American footprint, which entailed the closing of small Forward Operating Bases. One of these was Sabarat. One of the three S's was the first we would close (another S would close a few months later by the same crew and the final S would close a year later under the instructions of one of our crew members assigned to the air headquarters in Qatar).

Like every other day, our crew proceeded through our well-tuned rituals: wake-up, shower, shave, breakfast, brief, fly. When we walked into Ops after grabbing our survival and flight gear, the crew noticed that we had a Triple S. All three stops were to Sabarat. But the last flight of the day for us would also be the last US military fixed wing flight out of that small LZ.

Originally intended to become a major air base for the US military, Sabarat never grew out of its small shell and remained an LZ for most of its years as an American base. The plans to develop it into a major air base and nucleus of American power in the country never developed, and no aircraft bigger than a C-130 were ever able to enter the small FOB. For most of us on the crew, the base remained a small LZ from which we would often retrieve more equipment troops and cargo than we brought in.

Today was to be no different. After we went over the usual Intel brief, we studied how bare bones the FOB had become, and what that would entail. The small firetrucks that had been there were extracted by another C-130 and all that remained now were some Army troops, a few more pallets of cargo, and the forklift that was going to load those pallets onto the airplane. The plan was to go out, pick up the pallets, come back, then go back to Sabrat, pick up the last military forces and troops and return finally for the forklift (because it was a fat, heavy thing that for some reason, the military thought was worth paying thousands of dollars to keep).

The fact that there was nothing else meant that if there was a problem, it would be a big problem. If, for any unexpected reason, we crash landed (those are usually not expected), there would be no services there to help us. Such a crash would most likely block off the runway for any other airplane to come help, and there would be no firetrucks or other services there to help extinguish the burning wreck. There would also be no ambulance or medics nearby to help a wounded crew. Many would simply think to let it burn, but in Afghanistan, a large plume of dark smoke meant someone was in trouble – and unfortunately, that meant that some terrorist / insurgent group would want to exploit our newfound weakness. The small, token force, left at Sabarat was too small to help repel any serious attack, and the evacuation of the base was kept secret in order to help hide this weakness from the overall population.

Just to be sure though, we would taking with us our usual team of FAST (Fly-Away Security Team) or RAVEN security forces. These were Air Force Security Forces personnel that were highly trained in securing the aircraft they travelled in. Once we landed, they would swoop out and secure the immediate area while we worked on loading the aircraft. Normally, they would accompany us on various missions to other, dangerous LZs, but not usually to Sabarat. Given that we were evacuating and abandoning the base though, someone at headquarters thought it would probably be a good idea to give us some level of protection.

Fortunately, we didn't step to the airplane thinking anything would go wrong. Who does? Perhaps in arrogance, the crew walked to the plane like every other day, loaded the flight plan into the computers, checked the instruments, and left the back of the airplane completely empty. The stanchions were all brought down in the back, leaving the cargo bay ready to welcome the pallets for the first part of the mission.

SHANCS 55: Tetra, SHANCS 55, we're at 90YT4, Angels 9, headed to 88ER4.

TETRA: SHANCS 55, Tetra. Cleared direct 88ER4 at current altitude. Be advised, OVERWATCH 01, flight of two A-10s is above 88ER4, Angels block 10 to 12.

SHANCS 55: SHANCS 55 copies all. Cleared direct.

The flight of two A-10s were two air-to-ground attack fighters that could provide immediate assistance to ground forces if anything went wrong during the evacuation. They were to provide air support and stayed in orbit, high above Sabarat, as we flew low and below them to land on the small assault strip they were there to help protect.

After an uneventful flight to the LZ, we made a final left turn to line up with the zone. From the air and in the distance, we got a good glimpse at the base we were evacuating people from. It was largely empty of people and vehicles, but most of the structures remained. It was the small things that helped us realize it was really being abandoned: opened doors battering in the wind, small piece of trash here or there, but otherwise a sterility that showed no human activity was taking place anymore. The small FOB was nestled on the left side of the runway, and other than just a few tents housing the current forces left, there were 24 or so other tents, two large warehouse like buildings and that was it. The dirt roads didn't have any vehicles on them, and most of the containers we had seen in the past were also gone, slowly siphoned away by our crew and other C-130s over the past few months.

On our final turn, we saw a notification flash frantically on our instrument panel. It was from the airplane, casually telling us that a missile launch was detected. Our good propeller airplane reacted on its own though, and it automatically started to expel searing hot, burning flares in self-defense. On the crew's defensive systems panel, the flare count dropped quickly. A quick scan of the surrounding area, close to the runway, revealed no missile coming toward us. The crew looked at each other and decided to continue, rushing lower to the ground and the FOB's lonely runway strip.

Hammond again hit the zone on the tiny runway and the plane rolled to the back of the small strip where the forklift was waiting with the pallets to be loaded. Like all the other trips we had made to these tiny FOBs, the large ramp and cargo door were opened with the engines running. Seconds after the plane came to a complete stop, the forklift with the first pallet fought the engine exhausts and came up to the back of the plane to start the on-load.

A few minutes later, the last pallet was loaded and the forklift backed away as the ramp and door. Methodically running the checklists, the crew flipped the last few switches in preparation for take-off before pushing up the power and rolling down the runway to lift off. It all happened seamlessly and the few soldiers left on the ground only saw a C-130 close its door and immediately start to roll down the runway.

Forty minutes later, the plane landed again at The Hub, one third of the mission complete. Taking just enough time to refuel, remove the pallets, and re-rig the back of the plane to accept passengers, the crew took back off to repeat the same procedures again.

And repeat we did. Landing again at the bleak remnants of the FOB, the last group of passengers were brought aboard, along with a pallet of their bags they had been living out of for the past few days. When they were safely strapped down, the plane again raced back toward The Hub. This time, no flares were used and no shots against us were detected. To stay safe though, we again varied our departure like we had our arrival. Using different approaches and departures kept us unpredictable and optimistically, less of a target.

With our second shuttle flight complete, the crew again quickly reconfigured the airplane to accept a dozen or so passengers and the rest be a flat expanse for the forklift to park inside the plane. When that was complete, the engines were started again, and we nudged our way amongst the heavy flow of airplanes taking off and landing. We then settled down for the next forty minutes as we again took a different route to Sabarat, our last flight ever to the small LZ.

The plane landed for a third and final time at Sabarat. If everything went according to plan, no other plane would land here again. This was it, for the US military. As usual, we moved the plane to our "parking" spot, but in an unusual turn of events, the loadmaster got back on the intercom:

L: Yeah, we're going to need shut down. This is going to take some time.

The machine was so big that it was going to take time to load it onto the airplane without to make sure we didn't risk denting the airplane or accidently cutting off an important cable – like those that controlled the flight surfaces and the plane.

So the engine condition levers were pulled and all at once, the powerful motors whirred to silence. In the quiet silence of the barren, abandoned airfield, the crew slowly flowed out of the airplane and surrounded the enormous 10k AT forklift. The AT stands for Adverse-Terrain and its massive tires supported that claim. Alone, each tire stood at close to 5 feet high, and above them was the forklift that could carry ten thousand pounds. The AT forklift itself weight close to 33 thousand pounds. This wasn't the max weight the C-130 could handle; but it was getting close to its limit when facing such a small runway strip from which to take-off from, while carrying the fuel it needed to fly to the next closest air base.

Per the regulations and operating manuals, the AT forklift could fit inside the C-130. What had not been accounted for was the angle the forklift would have as it moved up the ramps to the cargo deck. Nor the height of the driver that would back the vehicle into the airplane. The loadmasters looked at the driver and started to discuss the hand signals. They would have to maneuver the vehicle, but also its hydraulics to adjust the angle of the lift-arm, and all this only after they let out some air from the tires. Feeling supremely useless, the front end crew started to instead focus on their take-off. Looking at the data about the forklift, they looked beyond its on-loading. They would have to take-off with this thing in the back. And it was *very* heavy.

McKay ran the numbers multiple times out of sheer concerns about the margin of error we were given (it was miniscule. There was no margin).

The loadmasters were still slowly, inch by inch, guiding the forklift back into the plane. The driver had to contort himself into unnatural positions to make sure he didn't get crushed by the plane's ceiling but also so he could still reach the proper levers on his forklift and see the loadmaster's hand-signals.

They would move back an inch before they had to lift the forklift's prongs a few inches higher and lower another piece to make sure nothing would scrape before then moving back another few inches. And so it continued.

In the meantime, the officers decided to golf with a metal rod they found on the abandoned airfield while the security forces we brought with us kept an eye out for anyone seeking trouble. Higher above us, the two A-10 fighters continued to orbit, also looking out for trouble. Clearly the officers placed a lot of trust in these defenders that they thought now was a good to play golf.

Finally! The forklift was loaded and the front end crew began to prepare for our departure. While we powered up the systems, Carter got on the radio and started to notify the appropriate agencies of our progress. This helped the

A-10s above figure out how long they had left (which mattered when someone considered fuel reserves something important). Helicopters also started to sweep in to pick up the last remaining soldiers that would be farmed out to other FOBs closer than The Hub.

The last security forces boarded the Army helicopters, but these remained on the ground, waiting for us to leave. They could pick up and go on a whim, but we were loaded with a baby-whale, which meant we needed to calculate our take-off power and requirements. Until we left, the helicopters would wait with the Army troops, just in case something went wrong and we needed help.

The forklift was supremely heavy. So heavy that we would have to burn fuel to make ourselves lighter before we could even think of taking-off on such a short strip. McKay ran the APU, the Auxiliary Power Unit, a small air-turbine generator that we used to power the engines, in order to start making a dent in our fuel reserves. The APU burned fuel much more conservatively than a jet engine, so while the APU ran, we started to run the data through charts and different books and computers to confirm all of our numbers. This would get the fuel burn started. Our main engines were also burning through the fuel.

Hammond, seeing the worry Carter had, highlighted that anxiety: "A healthy level of concern right now is not inappropriate," he said somberly.

Those words echoed further distress to the rest of the crew, and the Engineer checked the numbers *again*. In order to maximize our take-off performance, the winds would have to be in our favor as well, and they therefore had to be blowing from a specific direction. The only problem was that the airfield was nothing more than an abandoned runway right now. Except for two helicopters that were wondering why the heck we were taking so long.

Everything had to be taken into account, including the effects of wind on the take-off. Getting on the radios, the crew called one of the helicopters on a frequency that was common to all aircraft, and asked for winds. Most military aircraft have the ability to measure winds once airborne, and the Army Blackhawk helicopters were no different. Hovering a few feet above the ground, the helicopter relayed the winds it was analyzing back to us, and the Engineer was able to adjust his calculations. The winds were light enough and blowing close enough to the direction we needed that we were going to be within the airplanes' capabilities to take-off. We would be red-lining those capabilities, but we were within the performance envelope.

The forklift was finally secured and locked down in the back of the plane. It was now time to get moving. The airplane had turned around after it landed, which meant a small piece of the runway was behind us. It normally wasn't an issue, but we needed every foot we could get out of this flat piece of land. So we were going to back it up to the dried brush that was on the edge of our tiny patch of land we called a runway.

Putting the engines into full reverse, the airplane did not move. Under the immense thrust from the four massive fans that were the propellers, the heavy C-130 did not budge. After a few seconds of keeping the engines in max-reverse, the throttles were replaced to flight idle. The slope of the runway combined with the incredible weight of the airplane meant that we were not backing our way to the edge of the runway. We instead elected to go forward and turn around, head to the extreme back of the runway (and even beyond a little bit) to do another 180 degree turn. At the completion of this maneuver, the plane's nose wheel was placed up against the very edge of the runway, like a runner's hands on the thin starting line. The main gear was behind. But with the front, nose wheel kissing the edge, were technically on the runway, and that's all we cared about right now.

We waited a few more minutes for the fuel to get to levels that would make us "light" enough to take-off. And when those few minutes expired, while holding the brakes, the throttles were pushed to their max power.

E: 3 is good, 4 is good, engine 1 is good… A little more on 2. Two's good.

The engineer looked over the engine gauges to confirm that they were at their full potential.

E: Power set!

P: Ready, ready, hack!

Like other assault take-offs, a clock was started as the heavy, massive plane, slowly started to inch itself forward down the soon-to-be-abandoned runway. Slowly, but definitively, the plane's roaring engines accelerated the aircraft down the flat strip over the next few seconds, and we reached on time our critical take-off speed.

CP: Go! … Rotate!

Without hesitating, the pilot pulled on the yoke and the plane slowly rose into the air, the cliff-edge of the runway dropping below us as we got airborne and zoomed upwards.

Over the radio, an observant but unknown voice spoke up:

“Nicely done.”

*Yeah, it was!* thought the C-130 crew.

Clear of any obstacles in front of us, the plane's nose was lowered to prevent losing any more airspeed and risk stalling. Pitching for a stable speed, the pilot continued to allow the plane to slowly climb into the evening sky. The last fixed wing aircraft had left Sabarat. The airfield would no longer see US occupying forces like it had for the past decade or so. War, involving allied forces at least, no longer would be present in this neighborhood. The loadmasters certainly couldn't hear the cheers from the front over the loud engines that were pushed to the max, but they did smile to each other with a thumbs up.

As we flew away, we left behind an abandoned outpost from which the US military used to stage from. It was another sobering reminder of how the longest war in American history was slowly coming to a close. Forty minutes later, we landed at an airfield that was far from being closed, sneaking in between two landing aircraft and quickly parking at our usual spot. Landing at The Hub was absolutely no concern, given how long the smoothly paved runway was – the airport was used to seeing massive C-5 or 747s land regularly.

The engines were shut down and wound-down in thankful silence as the forklift was slowly removed from the cargo compartment. This was our last stop of the day so fortunately, we weren't in a rush to get it off.

Maintainers were here to help too, making the process a little more expedient. With more eyes looking at the critical points, adjustments could be made faster.

The crew returned to Ops like any other mission. The FOB may be closed and no longer a place for us to fly to, but we still had many more stops to travel to, and the war was not over for us as we checked the potential plan for the next mission we would fly. For the first time in many years, Sabrat was no longer on the schedule.

Instead, we would being flying alert-missions more often now that Sabrat was closed. Instead of the Army occupying a region, we would start carrying them wherever they needed to go, responding more like a police force than just sitting in someone's yard.

During the mission, two airmen from Central Command's public affairs office had been aboard the aircraft with us during the final evacuation to document the FOB's final evacuation.

## Realities of War

It's Christmas! We decide to break the rules and wear a different kind of uniform. Instigated by Hammond, everyone on the crew got ugly Christmas sweaters. So we wore our Christmas sweaters over our t-shirts instead of our super-warrior-like camouflaged uniform blouses. This was only to be done while aboard the airplane – we weren't looking for direct insubordination, but just a little rebellious Christmas spirit. We had been wearing the same uniform every single day for the past few hundred days. Every – single – day.

So we changed it up when on the aircraft, travelling across the warzone.

On this day, we were tapped for an aeromedical evacuation mission, which meant that there were wounded; people who had gotten sick or hurt while fighting in a war so far from home. We were going to be transporting them back to The Hub where they could get care from one of the best military hospitals in Afghanistan (if we could get the patients to The Hub, there was a 98% chance the patients would survive) or be flown to another base in Europe.

After the ritualistic briefings and pre-flight, the whole crew was at the airplane while the aeromedical crew prepared the cargo compartment for their mission. The aeromed team attached their litters and medical gear to the center stanchions, plugged other gear into the electrical outlets in the cargo compartment, turning the back of the plane into a mini-hospital.

The aircrew decided to change that. While waiting, we agreed seeing all this medical gear was not entirely within the Christmas spirit and we should do something to lighten the mood. We started to attach and string around the back of the airplane Christmas lights. Plugging them into the power units of the airplane, the aircraft interior lit up in a flood of multicolored Christmas spirit. With our holiday uniforms, we were going to be the next best airlifters on this day, after Santa, of course.

The plan called for us to do an around the country tour and pick up multiple patients as we went along. We would all bring them back to The Hub at the end of the day. Our first stop was Little Italy, where we were gaining popularity. While the medics were on-loading patients, airfield workers wanted to get a few pictures with the crew in our holiday sweaters, to celebrate Christmas with the weirdly dressed Americans. We even got a few thumbs up from the patients lying in their litters, glad to see the orange, red and blue Christmas lights strewn around the aircraft. The Christmas lights were also a big photo-magnet, as people took a few photographs to memorialize our decorations.

Leaving Little Italy, our next stop was Kandyland. After landing, more people wanted to get pictures with us, and we thought it would be nice to take a picture of all of us for families back at home. Carter would edit together a Christmas card to be emailed to the loved ones.

After picking up more patients from Kandyland, the final stop was the British Isle. Popping the hatch open, Carter, dressed in his Santa uniform, waved at the tower controllers as the plane taxied by. Sheppard and Hammond waved through their own windows and loads through the open ramp and door. The tower controllers congratulated us for making their holiday better over the radio and gave us a thumbs up from their high vantage point with beaming smiles.

Finally all loaded up, the plane took off to head back to The Hub, mission accomplished. The patients were all doing well and were about to be brought to The Hub's hospital.

Once on the ground back at The Hub, the 10.3 hour day was coming to a close. Until Ops notified us of what's called a refrag; an update to our mission. We weren't done just yet.

The Commander of ISAF needed a ride to the British Isle. And there were critical care patients that needed to be evacuated from there as well and brought back to The Hub. We still had enough time to complete the mission and we just came from there, so we were went back to the plane, ready to go. We were in fact mainly excited, although slightly nervous; nervous about having such a high ranking figure aboard, and we were starting to be late to pick him up from Gobble. The aeromedical teams had to swap out because of the critical care requirements, which took time, but in the end, it didn't really matter: they had the priority. The ISAF Commander would have to find another ride as we all waited on the plane for the medical teams to update their equipment and personnel for the new medical requirements, while the sun set on The Hub.

The Critical Care team was aboard and our mission was cleared to proceed directly back to the British Isle without stopping for the ISAF Commander. But in the process, we did find out why the Commander was moving to the British Isle at the same time as we were. As we rushed toward our destination, the medical team briefed us on the aspects of the mission. A Marine had been killed, and he was going to be moved back to the US. But he wasn't the only one that had been in the firefight that killed him. Other injured members had to be evacuated back to The Hub for critical care. Helicopters were on the battlefield evacuating them back to the British Isle, where we would relay them to bring them back to the Hub. Helicopters just didn't have the range, speed or tools aboard to get them so far (they were on the other side of the country).

The Christmas mood became very somber, and while en-route, we took down the Christmas lights and took off our Christmas shirts and sweaters. It was fun when all we were transporting were patients that were more stable and going home in a celebratory mood. That frame of mind had now vanished.

An hour later, we landed at the British Isle, where we parked and the critical care team rushed into the makeshift clinic to assess the patients. One of the medics came back out and claimed that they would need some time to stabilize the patients before they could be moved. Transporting patients in an aircraft can still be traumatic and even dangerous for their bodies. The medics had to first make sure they could be transported without making their conditions worse.

After relaying to tower that we were going to be staying put for some time, the airfield shut down for the Fallen Comrade Ceremony. We were the last plane that had arrived and we were the only ones who would have been permitted to leave. The airfield was shutdown, otherwise. The only other plane parked on the massive parking area was a single other C-130 with giant three-digit numbers located underneath the cockpit windows; this was the recognizable sign that it belonged to the Marines. They had been dispatched to pick up their fallen Marine and would take the fallen warrior back home. Our aircraft was the backup in case there was a problem.

While we waited near the aircraft in the cold December night, British and US Marines that were headed to the Fallen Comrade Ceremony invited us to take part in the loading ceremony, and we humbly accepted.

We walked together to another entry point onto the flight line, where we joined up with hundreds of other Marines and Soldiers who came to pay their last respects. We were told to line up into four lines, and we did. The air was getting colder every minute, and those of us that had brought our flight gloves with us put them on. Some of us were less lucky and had left them on the airplane, but given the reason why we were out here, no one complained.

After waiting a little more, we were marched out to the Marine C-130. The plane's ramp was open and a giant American flag was unfurled on the front bulkhead. The rest of the plane was completely empty. We lined up behind the airplane, two lines on each side of the ramp, leaving a space as wide as the ramp open as a path.

We stood at attention, in the night, waiting for the ambulance to arrive with the Marine aboard. And when it did, a Marine called everyone to present arms with a booming voice, breaking the silence that had prevailed so far. No one had been talking or joking around as was normally seen in large gatherings of waiting people.

At the end of the hundred man lines, the ambulance doors opened and six men took the casket. They slowly carried it toward the airplane, in front of every single man and woman saluting.

Slowly, the procession moved forward. It left time for all of us to realize that someone was not going to ever be home again for a Christmas. Someone had paid the ultimate price for this war. Our nation has been at war for over a decade – 10 years! That means that if that Marine was 20 years old when he died, he was only 10 when our leaders decided to go to war. A generation had grown up with war. The chances of us dying out here were still real. That Marine still decided to sign up.

The casket, slowly passing by in front of us, and the knowledge that more injured people who may not make it were reminders that no one planned to die. No one planned to end everything – memories, goals and ambitions, pet projects and relationships. Everyone always goes on assuming that you can always "do it later." But for that young man, passing by all of our salutes, there was no later. There was no consciousness anymore. All the goals he had set would never be accomplished. The planning he had done, the retirement funds he had started to place aside, the college funds for his future children... All of those were now finished. His plans to buy a new car when he returned from the deployment with the extra cash he saved up were not going to be used. And none of these addressed the plans others may have made with him after he returned. The parties he had been invited to, the dinner at a restaurant to celebrate his expected return.

As the casket was lowered onto the deck of the aircraft, the call to order arms was given, and we all dropped our salutes, remaining at attention.

Members of the fallen Marine's unit lined up behind the airplane to pay their respects individually over his casket. They walked inside the airplane's cargo hold and said a few quiet words. And after each of them said goodbye, we were released by the same Marine who had called us all to attention. There were no speeches or other distractions.

Silently, the hundreds of us that were on the parking ramp scattered away. There was no talking. No one was in the mood to talk. Even less joke around. Our crew simply, quietly headed back to a terminal next to the medical team's shack where they were treating the wounded. We still had to wait a little bit to make sure the patients were stable enough to be transferred to portable systems for their trip to The Hub. The medics had, during all this time, continued to work frantically to stabilize the other wounded.

Later, the time did come for us to evacuate these patients. As we stepped back to the airplane, the Marine C-130 took off over us and onboard, the fallen comrade. We entered our own dark airplane and McKay powered up the systems to prepare for the flight home.

About forty five minutes later, the patients were onloaded and we headed back to The Hub. Our callsign, MEDEVAC 55, gave us priority over almost every other flight around us.

After reaching our cruising altitude, we called Tetra:

<u>MEDEVAC 55:</u> Tetra, Merry Christmas, MEDEVAC 55, priority medevac leaving niner two hotel delta eight, destination five seven lima foxtrot two, Flight Level two three oh.

Tetra: MEDEVAC 55, cleared as requested.

But then as we continued to fly forward on our direct routing, we could hear Tetra do some extra work:

Tetra: Jedi 24, maintain two four oh, priority medevac will be passing beneath you, two three oh, single C-130.

JEDI 24: Copy that, we'll maintain two four oh.

Tetra: SPY 45, turn left fifteen degrees, vectors around priority medevac.

SPY 45: Fifteen left for SPY 45.

And these calls continued to be made through the night as we rushed back to The Hub.

Once we got closer, we got in touch with the air traffic controllers that controller The Hub's airspace.

MEDEVAC 55: Approach, MEDEVAC 55, priority medevac, level at twenty three thousand, request direct five mile final straight in visual, The Hub.

App: MEDEVAC 55, radar contact. Be advised, The Hub is under attack, the runway is closed.

The crew looked at each other. This was not very convenient. It was Christmas, so an attack was to be expected. Insurgents/ terrorists usually liked to pick important Western dates and anniversaries to attack our bases. September 11th was a *wonderful* day full of fireworks lobbed by terrorist. Christmas had already taken the life of a US Marine, and could cost more.

P: Medical chief, this is the pilot.

MCD: Go ahead?

P: How are the patients doing?

MCD: We'd like to get them on the ground as soon as possible.

We got back on the radio with approach:

MEDEVAC 55: Approach, MEDEVAC 55, copy. We're going to keep pressing forward and we'll talk to tower once we get closer.

App: MEDEVAC 55, copy. Cleared direct to a five mile final to The Hub, descend at pilots' discretion. Call airfield in sight.

Like Tetra did, the controllers parted the skies for us, getting other aircraft to move around us, stop their descents or ascents. We were the utmost priority in the skies. Perhaps it was because we were willing to fly straight to an airport that was being mortared and rocketed, or maybe it was simply procedure; perhaps a mix of both. Regardless, it was convenient.

While we continued racing toward the field under attack, the medical chief relayed the information through the flight crew to ground medical teams what they needed to know prepare for after we would have landed. The front end crew relayed it to the ground agencies. The ground agencies politely acknowledged the information by reminding us that the base was on full lockdown, being under attack.

Our crew acknowledged that fact, but we were still going to land.

MEDEVAC 55: Approach, MEDEVAC 55, field in sight.

App: MEDEVAC 55, contact tower, one two thee point four.

MEDEVAC 55: Switching to tower, twenty three point four, MEDEVAC 55. [pause] Tower, MEDEVAC 55, priority medevac, five miles, gear down.

Tower: MEDEVAC 55, be advised airfield under attack, cleared to land at your own risk. Men and equipment are off the active.

MEDEVAC 55: Ok, cleared to land, and we copy the airfield under attack.

We flew the flight like every other. The only difference this time was that we were the only ones landing. The base was closed to everyone else, and we were consequently making ourselves a nice target. On our traffic systems aboard the airplane, we could see dots representing other warplanes circling above The Hub, waiting for the attack to be over while we continued to dive down, toward the runway.

We waited to cross the base's perimeter before we turned on our landing lights and revealed ourselves. Helicopters were flying around, and we suspected they were the Quick Reaction Force deployed to defend the base. There was no way to know, but they were the only other air traffic that we had to watch out for. Far above us were other airplanes orbiting in the dark, waiting to their opportunity to land after the base would be declared safe again. You could see the other airplanes on the taxiways that stopped dead in their tracks the moment the attack happened. Fortunately, our path from the runway to the parking spot was clear and we proceeded to taxi to park.

After we parked, the all clear was given over the base's loudspeaker system, and the ground medical crews rushed to our airplane, ready to assist and evacuate the wounded to the hospital.

It was confirmed to us that this was the end of our day – we would not be heading out again.

As the medical bus rushed the patients away, our crew signed off the flight paperwork and maintenance logs, the maintainers looking wide-eyed after having seen the medics and their patients. The remaining medical crews dismantled their equipment and began to clean up and wash the back of the plane, with the front end crew soon joining them to help out.

After silently cleaning the airplane, we headed back to Ops after having dropped off our flight equipment at AFE.

We entered Ops and everything was normal. The sports channels were still on, blaring commercials about chips, various fast food chains and everyone smiling. Life was great. No mention was made about the mission. Some small talk about how the airplane behaved, but nothing about the attack, the wounded or the ceremony. It was part of our lives here, however subtle.

We headed to the DFAC hoping for some Christmas turkey, but it had closed, and we were left with the sandwich line. They weren't bad, and since they were grilled, it still counted as a warm meal.

When we got back to the rooms, a small red stalking was left at the foot of our door. It must have been there in the morning when we rushed to our mission, because it had been knocked sideways, and everyone else had already grabbed theirs. But it was the only gift to be received this Christmas. No wrapped gifts or gathering with the family. Such was the cost of being at war, as well.

In the back of our minds though, we couldn't help but remember the fallen Marine. We could complain all we wanted about missing Christmas with our family, missing our children's first gift unwrapping, or the last opportunity to celebrate Christmas with our very old grandfather. But we still had more to look forward to.

There was an often repeated adage that we signed up for this. The war had already been raging when we signed on the dotted line. The say was "You don't join the military and expect to avoid going to war." It's our job: the prosecution of war and whatever it takes to support those military objectives.

It unfortunately does not make witnessing the loss of life any easier. The belief and cause that made you sign up initially doesn't seem so rosy and beautiful when you sit eating a sandwich in a half empty cafeteria. It doesn't fill the void left by your missing family. Or by the void created in another family.

That's why you lean on your surrogate, however temporary, family. Your crew. You agree to do something special on Christmas, like getting sweaters and decorating your airplane. You wish everyone a Merry Christmas when you can. And you remember that you have plenty more to look forward to later on, and that just maybe, you did some good while out there. You did something that hopefully people won't argue or bicker over and affected a positive change for the greater good.

Wounded soldiers at The Hub's hospital were thinking those hopeful thoughts, as had been communicated to the author on a visit during a later deployment.

And hopefully, the friends and family of the late Marine also believed so.

## The Home Front

Having loved ones deployed can often be taxing for those remaining in the United States. There are the common understandings that a parent must now deal singlehandedly with all the children, bills and other upkeep of a house while the other parent is thousands of miles away. This seems fairly similar to the plight of millions of single parents across the world as well.

What was a real challenge for our friends and families that cared for us was the concern. The quiet anxiety that insidiously lurked in the back of our loved one's minds. For most of our crew, this was the first deployment we had ever been on – the first time we were exposed to the dangers of the area for an extended period of time. This meant it was also the first time our families and friends were dealing with us being there.

One individual had explained it through a different comparative analogy. The stressors and apprehension was similar to a parent worrying for their teenager the first time said teenager is late coming home at night, past their curfew. The parent starts to conjure up disaster scenarios, imagining the death of their teenage child etc. Irrationality starts to settle in. They consciously think about how it's just the teenager once again being late and behaving like a child more than anything. But the subconscious quietly pushes fear through the mind.

The same occurred for some that remained in the United States while our crew was flying combat operations over Afghanistan.

> "The difficult parts of the deployment came in small bursts. It wasn't a continuous worry. It's not like when you're waiting in a hospital waiting room to hear the verdict of a surgery. It was more episodic. It really became hard when I was alone. When I would ride home from the office or when I would lie in bed. It would come really when I was actually alone – not necessarily at the office when you're still surrounded by others. It was whenever I didn't have to keep my mind occupied. That's when I thought of you, obviously.
> That's when it would be come hard. I would wonder what you're up to. Sometimes, when you were gone for a few days at a time, or you

were 'overdue' for your check-in with me, that's when my imagination would start to conjure up crazy things. I'd remember the times you talked about the base being bombed and wonder if it had happened again and if this was the time you had been hit. I would wonder if the military had the means to call me. Or just contact me in general. I thought, well the Air Force would call your parents and then I'm sure they'd call me. It was like this dark black hole where my mind would just go stupid nuts and start making up ridiculous scenarios.

There were times I would just wait for my phone to ring, but that was the terrifying thing. I just didn't know if it would be you or if it would be someone with bad news. You know?

Not to mention that I did miss you. I mean, fortunately, we had already been apart before and for quite some time, but it just gets old to always be apart. To know that you were out *there*, instead of *here*, got pretty old too.

The slow Internet you had, if you even had Internet some days, was frustrating. I had to deal with work stuff too and sometimes just needed to talk to you and that's when you'd either be unavailable or you would have had a bad day too and we would end up fighting.

I guess that was part of the difficulty too. I just couldn't relate to what you were experiencing. I mean, when you said you had a medevac mission, I guess I could not understand what was happening in the back of your plane.

To be fair, you didn't really communicate it well with me either.

Not that what you want to do when you are talking to me is re-live your day. But the fact that I couldn't be there for you frustrated me, and I think it frustrated you, too. You trying not to worry me would end up having you bottle things up.

Then there were my parents. And your parents. They would ask me how I was doing, and I just felt compelled to tell them that I was doing fine when in reality I was really stressing or missing you. I didn't want to worry them.

[John] at the office would often have the TV playing the news, and there were times when they'd talk about some battle or whatever-attack taking place in Afghanistan, and I would just wonder if you were in it. I know, it's all irrational, but still – I guess I couldn't help but worry about you. I certainly didn't want to talk about it with the parental-units though.

I just missed you.

A lot.

I wanted to hold you and go live the idyllic life. I couldn't understand why the hell you were out there in the first place. I know you were doing some good things like delivering humanitarian supplies and helping some people get out away from bad situations, but there were days I just wanted to wake up next to you and go get a coffee at the shop next door, go for a walk in the park down the street or maybe go see a concert in the city. Not hear about how you were flying towards the gunfire to go rescue some guy.

So these occasional stressors, over months. It did start to become a burden.

I was so excited when you came home.

I'm really happy that that woman, your commander's wife, took a photo of us after you landed. I certainly am going to remember that moment for the rest of my life. Just like when I jumped into your arms after I saw you get off that plane.

That fat propeller airplane with a weird nose.

I'm certainly jealous you gave her more attention than me during those several months.

It was taxing and hard to know you were out there. My own ignorance of the dangers, my imagination, but the real dangers you also faced combined into this weird reality.

I'm just happy you're back."

While this was the honest feedback provided by one of the significant others of one of the crew members, the reality is that many of us missed home terribly. It had nothing to do with the physical home – it was the people.

The old cliché ringed ever more true while we were out the door: we don't realize what we have until it's gone.

But those we chose to be with and share our lives with were making an enormous personal sacrifice back at home, waiting for us to return. However this did not mean that the crew was not also thinking about their loved ones.

Re-reading old conversations from before the deployment would always bring smiles. One squadron member had a baby not long before the deployment and yearned to be with his baby boy again. He would spend time on his laptop, video chatting to see his little one, trying to be there as he grew up before his eyes, but thousands of miles away.

For those that were engaged or in a long-term relationship, the pain was sometimes difficult to bear as well. Granted, these are military officers and troops we're talking about here: they felt nothing and were trained war machines. So unfortunately, those back in the US wouldn't necessarily hear about the true feelings of their loved ones who were deployed. Pride was still a shell that hid secrets, especially so far from home.

But if you peeled back this façade, deep down you could see some of the crew dreaming of being back in a comfortable life with their loved one by their side. They would imagine going to a street-side coffee shop, worried about what time the theater started, basking in a 'normal' life: steady employment and hours, no long deployments to come between them, nor military moves or other military annoyances. Just a place they could settle down in and build roots and perhaps even a family.

They would think back to why they had fallen in love in the first place. They would remember the moment they had met, the laughs that had developed the relationship. The quirks and peculiarities, teasing and poking would force some to smile while lying in bed, remembering those who weren't with them while in Afghanistan.

Of course, again, they would never admit these things – that would be too 'mushy'; 'cheesy' – emotional. To remind our girlfriend what caused them to get a certain nickname was what could make a day easier. Or remember that a wife's distracted expressions could be so adorable. That their excitement and contagious laughs were what allowed us to feel *happy*. That their delicate love was truly what kept us going, even though we were perhaps naïve to realize it. Or a baby son's inattention to a camera except to try and reach the little green light would be the image that made you smile without explanation except perhaps defining love.

While couples still sometimes fought (or were having 'discussions'), there was no one else in the world they would rather fight with. Perhaps it was the fights that showed how deep their relationships were: it showed how comfortable they were with each other, and a willingness to be honest with each other, even under the circumstances they were in. That of course, didn't take the pain away, but it was during these fights that sometimes the true depths of hardship would sometimes reveal themselves.

During the deployment though, different conversations sometimes hid the concerns felt by those at home. Below are some email and instant-messaging chats from various crew members and some of their friends and family, displaying different attitudes throughout the deployment.

## The Cool Factor

# I can haz wingz?

From: Her
To: Him
Subject: I can haz wingz?

> Hi,
>
> Don't know if you'd go for this: but if I send you the big wings you gave me after you graduated flight school, can you send me yours that you fly with in Afghanistan? Or do you not have any "hard" wings with you in Afghanistan?
>
> PS. I love you.
>
> SugarBuzz

Delivered upon return to the United States:

## Worry

### Send me...

From: Her
To: Him
Subject: Send me...

...an I am still alive message, please.

From: Him
To: Her

Still Alive.
Unfortunately, the AF WiFi network is still not working for us...
I'm short turning to another flight in 6 hrs, so idk if I'll get to talk to you.
I love you,

Me

From: Her
To: Him

I hope you are sleeping at some point!

## Missing and Wishing We Could Talk

**Wife:** Babe
you should be flying,
but you liked the facebook comment
so you arent
come back
you shouldnt be showing as online when you aren't actually in front of your computer
do i go to sleep or do i wait for you
that is the question
most likely i will wait for youuuuuuuuu
hopefully you are just getting a nibble
**Husband:** Ok
Do you want to video chat?
I'm ready to go now if you are? I just have to walk to the WiFi spot.

## The relationship doubts

**Girlfriend:** hiii
i am about to go home - no internet until friday. are you there?
**Boyfriend:** Yes
I have just returned from flying
I work again in 9 hrs (non flying)
**Girlfriend:** how was it
**Boyfriend:** so I'm going to bed now anyways.
**Girlfriend:** ok
**Boyfriend:** I went to the German base again
and had

zer gut food

**Girlfriend:** happy to hear it

:)

**Boyfriend:** (mashed patatoes were AMAZZZING)

I do wish I knew how to speak that horrid language sometimes though.

**Girlfriend:** excuse me

you mean awesome language

you speak it more so than i ever realized!

**Boyfriend:** Yeah but I intersperse it with English, so therefore, it's not speaking: it's just wording.

**Girlfriend:** :)

genglish

**Boyfriend:** Englerman.

**Girlfriend:** ha! i like that

**Boyfriend:** But the German cafteria

**Girlfriend:** yes?

**Boyfriend:** WAY better than the American one we have here.

**Girlfriend:** nice

are you ignoring be btw? because that is what i have convinced myself of

**Boyfriend:** Why would I ignore you Sweetie.

I talked to you yesterday.

**Girlfriend:** i know

but ive just convinced myself that we reached some weird cliff this weekend and now we wont really talk anymore

**Boyfriend:** Sweetie

We shall talk more.

But right now,

you remember that I have a roomate

in the SAME room as me (not like you in the apartment)

**Girlfriend:** i know

but you know

so you are sure you are not actively making a point to talk to me less?

(i am sounding like a crazy lady - of this i am aware!)

**Boyfriend**: Sweetie
I am not making an active move to ignore you
not more than usual.
;)
And yes,
you are crazy: we established that at
**Girlfriend**: yep
**Boyfriend**: Ok
**Boyfriend**: So
Now that we've established that I'm not ignoring you
do you want me to call you?
**Girlfriend**: now?
you could
**Boyfriend**: Ok
Give me a sec.
Another sec sorry.
**Girlfriend**: Ok

## Reminiscing about the old days and missing your loved ones

**Girlfriend**: alo
**Boyfriend**: Hi
I'm back in my room now.
**Girlfriend**: you are still awake?
**Boyfriend**: Yes. You?
I was at the gym.
**Girlfriend**: i am still awake
I would hope so, at this time.
**Girlfriend**: you asked...
:)
**Boyfriend**: Sweetie
**Girlfriend**: mhmm
i told them today that you initially rejected my fb request
they found this quite amusing
**Boyfriend**: Who's they?
Your crew?

**Girlfriend:** chrstine
magge
and michael
so did you actually reject me and then i friended you again?
**Boyfriend:** Is that an actual question or is that how you told them?
**Girlfriend:** i am asking you
**Boyfriend:** Yes
That is how it went.
**Girlfriend:** how soon after your rejection did i refriend you?
**Boyfriend:** You friended me after our first class in Lead Gateway
and I regjected you and then you got mad and friended me again
**Girlfriend:** i actually got mad?
**Boyfriend:** and I already knew what you were like mad, so I did accepted.
Well you didn't think it was funny,
which means yes: you were mad.
**Girlfriend:** how soon after was that?
that doesnt sound like something id do ;)
**Boyfriend:** But you did. I wonder if Fb has records that go back that far.
PS: did you get my email.
**Girlfriend:** did you get mine
**Boyfriend:** Indeed, nice.
Our friendship is pretty funny.
**Girlfriend:** is it?
**Boyfriend:** Indeed.
**Girlfriend:** on Facebook?
**Boyfriend:** Yes
**Girlfriend:** i cant see any of your posts to me. just mine to you
which are very randow
**Boyfriend:** That's all I see too.
I see a few from me, but most are from you. You're a chatty cat.
**Girlfriend:** i guess
did you just not post anything?
lol
thats funny
**Boyfriend:** It wouldn't surprise me if I messaged it to you instead of wall it.
**Girlfriend:** probs

and then i always took it public

ha

**Boyfriend:** Indeed. How wude.

Anyways, Sweetie. I'm going to go to bed now.

**Girlfriend:** probably for julie's benefit :P

**Boyfriend:** Julie was actually cheating on me at the time.

But indeed.

**Girlfriend:** so sad

**Boyfriend:** Oh wells. She dumped the guy who took over me recently, so I'm good now.

I would have really hated to be replaced permanently by THAT dude.

Bc I was clearly better.

It appears that I have not paid rent for the past two months.

Therefore, I must now pay it.

**Girlfriend:** oh yikes.

**Girlfriend:** amanda might be thinking about giving up the apartment completely

...

**Boyfriend:** ?!?!

Well she'd still have to find a replacement.

**Girlfriend:** yeah

**Boyfriend:** Or you can crash at my place for a while.

**Girlfriend:** :)

i do have a key

and can drive the car out of the garage

**Boyfriend:** Indeed. It's time for bed now

You're busy at work anyways.

So I'll ttyt.

**Girlfriend:** okidoki

you are doign a good job not making me think you are ignoring me :)

**Boyfriend:** Ja this is good.

**Girlfriend:** ja

indeed

**Boyfriend:** Alright. Good night Sweetie.

Bye bye

**Girlfriend:** good nighty nights

## Frustrated We Can't Talk

**Her:** hi

i was doing laundry/locking myself out

want to video chat?

alright. going to get my laundry

**Him:** Nope. It's too late. I'm not getting dressed again to go outside and video chat.

**Her:** hi

**Him:** Hi

**Her:** Sorry I missed you before

I'd locked myself out :)

**Him:** Cool

Well, Im going to bed.

Glad to see you're back in your place.

**Her:** Ok. Sleep well.

Hope you had a good day. Don't be too grumpy :)

**Him:** K

Next time, when I try calling you on a variety of types of devices to video chat with you, answer.

Especially if I'm in a war-zone.

**Her:** Ok.

**Him:** Wouldn't want to regret that, would we?

**Her:** We wouldn't.

I wasnt planning on locking myself out. So I didnt take my phone with me :)

Sleep tight

**Him:** I'd love to know when you plan on locking yourself out.

Good night.

**Her:** Bye, darling!

## Helping Those at Home

**She**: alo
**He**: Hai
**She**: :)
**He**: Whatevs.
    Anyways
**She**: anyways
    grum grum grump
**He**: Honey
    I'm sitting "alert" this week
    which is just like back home
    I wait for a call.
**She**: harumpf
    thats boring
    thats what you are doing all week?
    gym time!
    i i i work outttttt
**He**: You're going to the gym now???
    What happened to work?
**She**: no
    you should go
    since you dont have anything to do
    make your triangle a square
**He**: I have been making it a square.
    the issue is that to go to the gym, I have to wear body armor on top of my PT clothes. Which I def don't want to do.
**She**: wow
    was it like that when you first got there?
**He**: Yes.
**She**: why?
**He**: Because the military is worried someone might start shooting fireworks and those could land accidently on base.
**She**: oh
**He**: Yes.
    So

I got my camera fixed btw.

So we can video chat properly again.

**She:** oh sweet

i cancelled celli today

so ill be home

i am getting really worried about my phd apps

because i dont think germany will pan out. its already been 2 weeks now and i havent heard anything

when i talked to my dad yesterday, he said that he is sure i didnt get the position

since its been so long and i havent heard anything from them

**He:** That's annoying.

**He:** But don't worry sweetie. You can always try again elsewhere within the American empire.

**She** :annoying isnt the word i'd use

but it is dissappointing

for sure

True. It's just that if I'd gotten that, my parents would feel like I'd made it, which would be nice :)

**He:** Honey

Your parents already think you made it.

You haven't gotten pregnant before getting a job,

so you're good.

And you work at NORTHWESTERN !

**She:** They don't..

**He:** Don't worry. If I'm proud of you

your parents are proud of you.

**She:** Nope, my dad at least views it as me biding my time until I get a PhD position.

**He:** Even if that's the case, then that means he believes you are on the right path.

And therefore is proud of you and believes you will make it.

**She:** :)

**He:** So Sweetie

The only thing I think you could do is earn more moneh

then I'd probs quit the USAF and del with your insufferable attitude.

**She:** That is not how my parents see it...but yes, in the normal world, this is true.

**He:** How do your parents view it?

**She:** Well, my parents first of all think that what I am currently doing is just a way to pass time, and that the content won't pan out to be worth anything.

If i get a phd positon in the US I will have wasted my time in the netherlands because my phd will still take 5 years regardless.

**He:** Your masters wasn't worthless bc it's what got your position at NWU.

That, and it saved you from living through flight school.

**She:** And all in all, the content of what I have chosen to pursue is murkey at best.

**He:** Do your parents know that? Or think that? Or do you think that?

**She:** I think that they think that :)

**He:** Oh, well you also think that I'm going to die here, and that you know it all, and that you are right about everything. So we know that what you think is as far from the truth as you can get. Which means, as I stated before, your parents think two thumbs up of you.

Four, actually, bc your parents both still have two hands.

**She:** Well who knows.

**He:** Well, I think that your thought process is a lot closer to mine than perhaps you like to admit.

You resent me for trying to impress ppl with what I do

but you're doing the same for your rentals.

**She:** I would say that I am not trying to impress, just looking for approval.

**He:** Well those two are synonymous. If you don't approve, your certainly not impressing.

But i'm not worried about my parents yet: i'm still trying to get YOU to approve and be impressed.

**She:** Impressed is different

I definitely do not think what you are doing is worthless.

but maybe I give you the same impression that my parents give me of my choices which would be sad

**He:** What do you mean?

**She:** Well if I convey upon you the same feeling of luke-warm approval of your path that I perceive from my parents, that is sad.

**He:** Hmmm. Well, don't worry Sweetie.

Since you're never impressed, it always seems like you think any idiot could do my job. Which I'm not truly fond of since I don't want to be an idiot.

**He:** And don't forget: you are often very much like Joe.

**She:** Hm.

well.

she is in germz right now

**He:** Ah

Well I was in what I like to call Little Germany yesterday.

**She:** The German Base?

**He:** Correct.

Das German controllers are very awesome to listen to.

And I got you something.

**She:** :)

it cost you a coin!

**He:** Yes.

We don't wear patches when we fly, so I had nothing else to trade him for.

**She:** ohh

what did you say?

**He:** I asked if there was a shop that was open for me to buy a coin or patch, and he

accepted to trade.

Bc at 11 at night, everything was closed unfortunately.

Including the German cafeteria, which is my favorite in the country so far.

**She:** I SEE

**She:** nice

**He:** He's a driver of what we call a "follow me" truck, because we follow it to our parking spot.

And I was helping him plug in a power generator into the airplane when it came up.

**She:** gotcha

**He:** So

**She:** that pciture is amazing

**He:** Thanks sweetie.

Well Im going to go now.

**She:** ok
**He:** I guess I'll ttyt.
Good day.
**She:** good day.

## Trying to Normalize Life

**Girl:** Honey, do you want me to buy you anything from the gap? i get $75 off if i spend $150. ex: http://www.gap.com/browse/product.do?uhoh&vid=1&pid
**Boy:** Word. Make me look preppy like that guy.
If you can/ want to.
**Girl:** do you think you are a medium or large?
**Boy:** i think large
you dont want it to be too short
I am a rather rotund man.
**Girl:** you arent
you are perfect :D
**Boy:** Oh sweetheart
I just discovered a song from Jason Mraz that I really like: Butterfly.
And I think you might like hte title.
**Girl:** do you want the grey or blue?
if you have a choice
**Boy:** Blue
Cuz it's the Air Force blue.
You just loled.
**Girl:** HA
this is a dirty song!
**Boy:** -_-
Now I have to listen to it again.
from Jason Mraz?
He's not the type though.
**Girl:** yer
**Boy:** Give me a second.
And yes indeed - he's a pole and she's only wearing shoes.
I can hear it indeed.
Oh wells. It's still a good beat.

**Girl:** lol
my new fave song
http://www.youtube.com/watch?=1qw9QQ8-Q
this whole new blunt album is da bomb
**Boy:** And this is pretty cool toon.
http://www.com/nottherealurl/
I sometimes miss France.
**Girl:** oui oui
je vais dormis
je suis une poete and didnt even know it
je vais dormir*
going to bed
<3
**Boy:** Ok
Good night.
And you are not a poet, because you don't know what rhymes.
I do.
**Girl:** ha
ok
night night
**Boy:** Gnight.

## More Worry

## lifesign

From: Her
To: Him
Subject: lifesign

please!

From: Him
To: Her

Sign.

## Family or Friends Thinking of You

### Some News

From: Aunt
To: Nephew
Subject: Some News

Daniel,

A small word to tell you that we are thinking of you. I don't know where you are at this time, but I hope that it's not too hard. I know through your father that you will not be coming to the village over Christmas nor at the birthday for Grandma and Grandpa. We will miss you. We'll send you pictures! Elice has started working again in anticipation of the entrance exams in January, so I don't know if she'll be coming to the village either. William has changed high schools and he is delighted.

Daniel, we are fondly thinking of you, please be careful and take care of yourself.

## You Thinking of Family or Friends

### Happy Birthday!

From: Me
To: Friend
Subject: Happy Birthday!

Hey man,

Sorry for being the same as everyone else with just the simple message, but I just wanted to wish you a happy birthday! I'm out in Afghanistan right now which is why I'm not being a better person and calling you, but rest assured: you will be annoyed soon enough.

I hope things are going well in NM, and I fully expect Anita to prepare a full fiesta for you. Say hi to her for me, of course.

Feliz birthday (my Spanish is slowly improving!)

From: Friend
To: Me

Hey Bro!!!

I had not idea you were across the pond...that is just AWESOME!! I bet you are having a blast. I hope. Thanks for the birthday wishes and no worries bro I'm modern, I don't mind email or texts lol. Anita is doing good...no big fiesta this year just went out to eat at a local brewery (the only one for 500 miles), but it was a great time. I'll will let Anita know and I'm sure she will say hi back...so hi from Anita.

Anyways, great to hear from ya let me know if there is anything you need that I can send to you. Take care and keep having fun.

Hasta luego,

Soto

## Missing Birthdays

# Happy Birthday

From: Brother
To: Brother
Subject: Happy Birthday

Happy birthday Daniel ! Everyone is thinking of you and misses you a lot ! Good luck! Don't die please. That would really suck. Really suck!

## Catching Up

# (no subject)

From: Me
To: Colleague
Subject: (no subject)

Hey Balthazar,

As you can see from our e-card from us to you, we got your awesome package. We're planning to watch [the film you sent us'] on Christmas night, just to understand what a cool crew does according to that film. And just to annoy Mr. McKay, I use your little Star Wars keychain to have R2 respond to my checklist items (R2 is after all, Skywalker's copilot and nav). I also steal his Rip Its, lock up his seatbelts, and tell him to hurry up and shut down the APU after we've landed so I can go smoke his cigarettes. Then he gets back at me by turning off the power to the fuel gauges so I think we're out of fuel... It's a welcome source of fun out here.

Things out here have changed a lot; I don't know if I'm making it up since our crew has drastically changed, but there are some things that I can't make up. As I mentioned before, the room setup has completely changed. Enlisted have moved to another dorm and are tripled up, and us 3 officers were also tripled. Habits and leadership styles are also different, but we're adapting.

You've probably heard of the Force Shaping plans the USAF has for us; I know a certain Major whom I'm rooming with is seriously considering taking the VSP. Have you thought of what you might do? I know you just re-upped, but that was before the Air Force offered to pay you to get out. Aiden and I are just in that bracket where if we get kicked out, we get nothing, but we're not shielded from the cuts either. We're both kind of stressing out too because we've just been asked by Col. XXXXX for an ADP, even though we haven't been on station for even a year yet. The rumors that the TFIs are all shutting down are things we're not looking forward to face when we get back from [The Hub] and need to plan our futures.

Here I am complaining, but you've got it also annoying, with the prospect of another 120 day deployment to [Glitz]. I suspect you must be fairly frustrated. I hope your wife is doing ok with the prospect of you being gone again. Is her father doing a little better? [...]

You'll be happy to know that the heating system on our side of the building is malfunctioning, which means I now live in the same igloo you and I lived in at the start of our deployment. I'm wearing socks and two blankets - blasphemous. I hope you enjoy your delicious warm food without having to don IBA to go to your kitchen.

[Live long and prosper,]

Aiden

## Celebrations

## RE: Captain Selects....WOOOOAHHHHH!

From : Girlfriend
To: Boyfriend

YAYYYY

THAT IS VERY EXCITING. Congratulations. I knew this would happen which is why I bought ingredients for a treat package today.

CONGRATULATIONSSSSSSSSSSSSSSSSSSSSSSSSSSS

-buzz

At 9:05 PM, ‹ XXXXX.XXXXX@XXXXX.af.mil› wrote:
-----Original Message-----

From: XXXXX /CC
Sent: Wednesday, 5:21 PM
To: XXX EAS All
Subject: Captain Selects....WOOOOAHHHHH!
Importance: High

Weasels,
Please congratulate the following XXX EAS officers who've been selected for promotion to Captain:

Aiden Sheppard
Daniel Carter
XXXXX XXXXX
XXXXX XXXXX
XXXXX XXXXX

This is an important milestone....for the reasons 19 AW/CC Col XXXXX illuminates below. Please have a read whether you're a Captain-select or teammate!

XXXXX XXXXX, Lt Col, USAF
XXX EAS/CC
DSN: XXX-XXX-XXXX
VoSIP: XXX-XXX-XXXX

> Captain-selects,
>
> Let me remind you selection for promotion is a recognition of future potential, not a reward for previous work. Captain is the rank where great things are achieved at the unit level. Captain is the rank where in most career fields we add the word "commander" to your duty title - as in "flight commander" or "aircraft commander" for flyers. It is the first rank where we ask you to not only apply your technical skills, but use your judgment to lead others.

There are about 330 million people in the United States, and approximately 330,000 people in the U.S. Air Force. That's less than one percent of the population charged with defending its air, space, and cyberspace. Of those 330,000, approximately 65,000 are officers, with 22,000 in the grade of captain. Indeed, captains comprise the flying force of our Air Force.

Our history is full of famous captains. You are fortunate enough to wear the same rank as they did, remind yourself of what they accomplished and the limitless horizon before you as you make your own mark on our Air Force. Remember, every promotion is an honor that comes with great responsibility.

Good luck ... captains

VR,
Col XXXXX

//signed//
XXXXX XXXXX, Colonel, USAF
Commander, 19th Airlift Wing
Little Rock AFB, Arkansas

## More Family or Friends Thinking of You

## Hello

From : Mother
To : Son

Hello sweetheart,

You are the hero of the office: everyone knows that you are in Afghanistan and asking for news, I am very lucky and certainly appreciate it. Grandpa and your aunt are thinking of you a great deal, and Grandpa is so happy for you that you are at the heart of the action. I am also very happy for you, even if from time to time as a mother, I can't forget the first time I saw you and I had this indescribable feeling to want to protect you...

Life moves on, and it will soon be your turn to undergo this for your own children, you'll see, it's wonderful.

Here, things have changed in the apartment. Alex has sold his house in Virginia and is waiting to move into the house in DC. Charles has to adapt to a new house where things don't exclusively revolve around him... Alex was at the XXX XXXX when the killer took action. He's fine, but a close colleague of his lost a few employees in the massacre. He didn't want to go the memorial service with president Obama this morning.

Friday, we went to a retirement ceremony for a Navy officer from his office at Annapolis, and it was very interesting. I of course thought a lot about you and about what kind of homage you would receive when it will be your turn, I'm sure it will make me cry with emotions.

In a simple word, I'm thinking of you, I hope everything is going well and that your work is giving you the satisfaction you were hoping for.

Affectionately,

Mom

## The Return is Happening

### Numbers JIC

From: Him
To: Her
Subject: Numbers JIC

In case Security Forces doesn't let you in for whatever reason, even with your ID, you can call the SDO Desk (123.456.7890) and tell them you're coming to welcome home a deployer, and ask the SDO if they could give you a ride from the gate to the Fire Station (that's where the families and friends will be waiting).

Otherwise, here is Capt. XXXX's number (we had dinner with him and his wife at the Castle at that fish place): 098.765.4321.

From: Her
To: Him

Thanks :D

SEE YOU TOMORRRRRRRROW

## Returned

### Hey

From: Me
To: Military Mentor
Subject: Hey

Hey Col. XXXX,

Have you moved to the Mile High City after all? I thought I remembered you said something about a possible airline hiring that would get you up here in the Denver area, but I can't remember exactly what the conditions were?

I (and XXXXXX) just got back from Afghanistan back to our respective home stations, and it's a sort of bitter sweet return. While it's nice to be home, flying in Afghanistan was a lot of fun. Now all we have to look forward to are checkrides and SKE to Vis LL to Pro ARAs etc. And the usual pesky side jobs and volunteer requirements.

Afghanistan was a lot of fun, flying ops and airdrops. It was nice to finally see the culmination of almost 2 years of training come together (although I'm convinced training did not have to take *that* long). My crew [...] was simply awesome (my Aircraft Commander was a Maj. Hammond who had been at LRF a while ago before he became an AMLO). I can't remember if you had ever deployed to The Hub, but if you had, I'm told it hasn't changed at all. We lived in containers which I thought were pretty nice, but still had to spend the night in the Herc a few times, once after one our engines starting to torch at another airfield, among other events.

How about you? Are you still racking up the hours to rejoin the airlines? I should be working on building hours myself, but ski might come in the way of that. Otherwise, how are you looking with the changes coming to the USAF (the 25,000 cuts coming down) and the shifts within the 130 community (still looking at being operational)? I'm losing track of all the changes happening; the Little Rock Wing CC apparently told our unit that TFIs were about to be shut-down this year. I wasn't there for that briefing (I was in Afghanistan) but that's what the rumors have told me from home station.

If you are in the Denver area, let me know and we can maybe meet up for a lunch or something.

X.X.

From: Military Mentor
To: Me

Good evening,

It's good to hear from you! Sorry for my delayed response, I haven't had the opportunity to sit down and respond as I'm always on the run. I didn't pursue flying for the airlines which could have possibly taken me to Denver. I made the decision the Regional airline life wasn't for me. I have two and half years left until I'm eligible for retirement so the plan is to finish that here in Little Rock. I'll fly with the Reserves when I have to and/or need to but my main focus is flying civilian at KLIT. I ended up getting hired to fly charter at Central and I enjoy it....I'm flying a Beechcraft Baron, a Beechcraft 400 as co-pilot and they're going to train me to fly King Air 90's and 200's. Good to hear that you enjoyed flying during your deployment and doing airdrops. I flew to The Hub a lot but never deployed there. When I was operational in the C-130 we deployed to Glitz, Glamour and the Transit Center. Glad to hear you had a good crew, at least during the first half of your deployment....people always make or break your experiences. Skiing sounds fun, you're based a in a great location for that...take advantage of it! As for the downsizing in the Air Force, we're really not seeing that in the Reserves yet and luckily our C-130's are safe for now. Currently my unit is transitioning to a Reserve operational unit so we'll be training to deploy, not sure if I'll ever have to go though. The ANG now has the FTU mission. Are you eligible, given your year group, to leave the Air Force given the AF's intent to downsize? Enjoy your downtime and Happy New Year to you as well! Keep in touch.

## Leaving

As the deployment reached its end, many in the unit had hit a psychological wall and were living the same day over and over again. This was normal for any deployment. Our crew had been fortunate to continue receiving exciting missions and with the holidays to break up the monotony, the feeling of reliving the same experiences was limited. It always felt new and exciting to receive our orders and accomplish the missions tasked. But the excitement was also palpable when looking at the calendars. Some on the crew had hung calendars in our container rooms and X'd off each day as they lived on day by day. And the number of white spaces remaining before the "home" block was shrinking. Rapidly. When we could see the X of the day and the "home" day on the same page, we knew you were getting close to the end. For some, the minds of those who hung up those calendars had already jumped to the last block, but for most, they kept performing their mission as well as they had throughout the surge of operations.

The biggest changes we had to face were changes in accommodations. Some other units on the installation were also transiting to new aircraft or members, and space was becoming scarce. The order came down that we had to triple-up in our rooms. The cramped quarters we had were now padded with an extra 6 bags and a third muscular military officer. Whatever free space was left in the room was simply used to move from the door to the bunk bed. There was no longer any room for the personal furniture we had built. We had to change rooms as well – different areas of the container kingdom was being cordoned off for a larger group of women and they took over the area we had been in before. This was not bad, for more than one reason. One of these being that it forced us to pack our gear, which gave us a head start over the inevitable duty that was to come in a few weeks.

But for now, the countdown to our departure became ever more crucial as we struggled for private space and time. We flew and lived together for months on end, and the little shred of privacy we had was now taken out. Different hangout spots were found around the base; abandoned tents, morale centers or even Ops, were some of the few options we had to break off from each other and speak to our loved ones privately. Otherwise, there was almost always someone in the room. Rooms we would abandon in a few weeks as well, before the end of our time at The Hub and this deployment. This move only preceded another one before going home.

When we had arrived, we had gotten our rooms on the first night here. This was a luxury we intended to pass down. After all, if the tradition held, that would mean on our next combat deployment we would be allowed to sleep in our rooms on the first night too. It wasn't so much selfless as it was selfish. All of us were not disillusioned enough to think that this was the last time we were deploying. We were only hoping it was.

Regardless, it meant that about two weeks prior to our departure, we were moving into tents to free up the containers for our replacements. Contrary to the tents we had at the Transition Hub, where aircrew were given plywood walls for a semblance of (but ultimately nonexistent) privacy, these tens were simple mass sleeping sites. The long arched, sand-colored tent housed 25 bunk beds, squeezed inside like a jigsaw puzzle. That meant 50 people could sleep inside, and that wasn't enough. Another tent had to be released for some extra crewmembers to sleep in. The only thing separating our bunk beds were the small lockers we were assigned. But otherwise, our private space was now our mattress and the space below it. If you were on the top bunk, you had the top of the locker. Those on the bottom bunk also rigged bed sheets around bunk-bed frame to create a private, tiny yet comforting bedroom. It did absolutely nothing to keep out noises and even shadows could be seen through, but at least it helped to fall asleep without staring into the eyes of the man next to you.

Because all aircrew slept in the tent now, and we were still flying missions like any other day, it meant we had to be very cautious whenever we moved around. While it may be the middle of the day for our crew, it was the middle of the night for the other crew. Patience was tested more than once, and silence was never so missed. The only entry and exit point to the tent was a simple flappy door. That meant that if you had night missions and slept during the day, anyone entering or exiting the tent would also allow an ocean of Sunlight to come in with them into the tent. This was in stark juxtaposition to the constant state of darkness the tent was kept in to allow for any sort of sleep. We lived with dimmed flashlights in our mouths to keep both our hands free while we were searching for the next pair of underwear we had to bring to the shower with us.

These showers were, of course, in a separate trailer which we had to walk to. Other than the gym, it was the only site on the installation we were permitted to travel to without our weapons. And there were no bunkers between the tent and the latrines so if we were attacked, we would have to dodge into a nearby building.

Being in the tent during an attack was as dangerous as being outside. The thin piece of cloth that shielded us from the wind and rain did absolutely nothing against a falling bomb nor the blaring sirens and voice announcements revealing an attack was occurring. These announcements were useful for others, but for us inside the tent, they were redundant. The explosions and whistles were clearly loud enough as we stayed on the ground, huddled inside the tent. Those sleeping became even more grouchy when they could hear a whole attack while they should have been sleeping. They were staring at their watches not for the attack to end, but for their replacements to show up so that they could finally get a good night's sleep in a soundproofed building.

Until then though, the operations and missions continued to drop. We had to deliver arriving troops to far-flung FOBs, resupply others, look at airdrops, and plan for trips outside of the combat zone. The crews continued to work on their responsibilities back in the United States through their email networks, and families still had to struggle with the fact that even though the return date was so near, it still wasn't here. The stressors of having to face job cuts and potential job losses were not helpful either. The rumors of units closing spread as quickly as morning sunlight, and other rumors about potentially extending the deployment fed off that fear. One well founded rumor could spawn countless others that were completely absurd. The only way to destroy them all was to leave.

Finally, that day did come. The replacements arrived and slowly, one by one, our crews were released from operational missions to finish departure checklists. Like we had received when we arrived, we gave each new crew an additional officer from our crew to give them their introductory flight. The contrast from our first flight and our last was like night and day.

The jamming was no longer a problem for us, virtually non-existent in our attuned ears. We were fluent in the coded languages of the combat zone and we no longer felt compelled to wake up hours before our flight out of stress to prepare. We were prepared and could hop onto any aircraft and get any mission done. Such was the benefit of being the senior – the veteran.

It hit us then.

We were *veterans.*

Veterans of an actual conflict. Combat veterans. For years the only image of veterans that came to mind were that of old World War II seniors and their black caps embroidered with golden letters. We weren't wearing caps (yet) but we were veterans.

The new crew struggled with the radios as we had, but in reward for their good landings, some of us brought them to the hidden local pizzerias and told them at what time the delicious German cafeteria was open. We could spot problems before they ever became problems and eased the new team into their new responsibilities. They were going to be picking up the ball like we had months before. Headquarters wasn't going to slow down its demands or missions because of a bunch of new guys showing up.

At the conclusion of the introductory flight, the crew that was being replaced was signed off and was allowed to continue their out-processing journey. They signed out of communications centers, had accounts wiped. They turned in their weapons to the armory, passed customs, got financial paperwork in order and then just *left*. As one replacement aircraft arrived, a veteran plane departed.

But to add to the confusion, the hard crews used for our deployment were broken up. Those with wives and children back at home were given slots on the first plane out of Afghanistan. Everyone that could fit on that airplane would go. Only a minimal presence of maintainers and the leadership package would be on the last airplane out, slated to exit the country 24 hours after the first plane. Our new crew had to be the last one out. This was simply due to the fact that the commander was from our unit in Clearwater and he was to be the last one to go. All the other squadrons were permitted to return as soon as they were replaced.

Regardless, the transition only took a few days which wasn't a terrible problem. We feared our replacements would delay due to maintenance or other issues, but they were good team players and showed up on time. So we followed all the other crews through the different pit stops to sign our paperwork, take out our trash and provide our bags to the 'port-dawgs' (aircraft cargo riggers) to be rigged onto our airplane. Only a few of the original crew remained for the last flight out. Everyone else on that flight was a mashup of qualified aviators that had worked in leadership positions or other crewmembers that had arrived in the Middle East later on in the deployment. But this was part of the excitement of leaving.

A pair of U.S. Customs agents met us in a hanger to ensure we emigrated properly out of Afghanistan, before we signed the paperwork signaling we were no longer assigned to our combat, deployed unit. We were transiting back to the United States to be reassigned to our original home-unit. Showing up at Ops for our final flight brief was very odd.

Around us, the members of the squadron in the Ops building had new faces. Our personalized decorations on the plywood walls of what had been our office area for the past few months were taken down, the tail numbers and names of planes and crews on the blackboards were changed with new aircraft numbers and names, and the briefers were the new arrivals. We were the veterans of the current war – we could have practically given the brief. This was an arrogant feeling that stayed silent of course. While we were capable flyers now, the shock to our systems all of us had gotten when we first arrived a few months ago had still taught us a level of humility.

As we walked out to the airplane, we grabbed our gear for the last time. All our bags had been loaded onto a pallet on the aircraft, and our maintenance friends were the only passengers for the day. No strangers on the aircraft today. Only strange airplanes around us as we walked to our last familiar tail which was to be our home for the next few days on our transit back to the United States. The rest of the aircraft on the flight line were the new jets that had arrived to replace us. The planes to which we had developed an attachment were all gone, repositioning themselves elsewhere in the world. Some had a problem with their yaw-trim motors, while some had a bad gauge on the number 4 engine TIT… These new planes would have their own quirks, but we weren't going to learn about them. We were leaving.

We started engines and backed-out of our fortified parking spot. Everything had a very "last-time" feeling to it – our next parking spot would not have T-Walls surrounding it. We wouldn't be running the combat checklists the next time we flew, and we were no longer reporting to Headquarters for this combat zone. We already had permission to take-off because we weren't supporting the war effort anymore – we were just flying combat support missions and returning to the US. Instead, we were assigned to a different command center whose responsibility was to get us home. This was the last time we would be talking to Ground controllers, Air Traffic Control Tower, and the last time we would be requesting take-off (all of those silently preceded with an enormous "hopefully").

<u>GRASP 02:</u> Tower, GRASP 02, holding short at Hotel.
<u>Tower:</u> GRASP 02, on departure maintain runway heading until 3 DME then left turn to 180. Winds calm, runway 03, cleared for take-off.

We lined up on the runway and punched the throttles forward, speeding down the runway of The Hub for the last time.

# Return Home

## Phase I – Escape

As we climbed away from the ground, and turned toward our heading, we saw what had been our home for the past few months slowly get smaller out of the cockpit windows. As we leveled off, the usual checklists were run and completed. But once the normal checklists were completed, we all silently opened our checklists to a new page that was a little different in anticipation of what came next: the Combat Exit Checklist.

The decision had been made that if anything went wrong before we exited the combat airspace, we would turn around and return to The Hub. Anything beyond the border, we would take the problem with us to our next destination. Our flight plans had us stopping in Syldavia again, then Scotland and Canada before reaching our home airport. But until we passed the border point, The Hub was our fallback.

We flew at over 300 knots over the ground toward our point of no return. The escape point. That point, sitting on the edge of Afghanistan was our goal. But that didn't mean we flew the mission any differently. It was over an hour away, and our day had started off slightly differently than usual, after all. We had packed all our personal bags into a pallet the night before, for that pallet to then be loaded onto the airplane. We still had fitfully slept in a tent (insurgents had launched an offensive against the base during the night, launching rockets and waking us up instantaneously with their ruckus). We had skipped breakfast to finish our out-processing paperwork and attempt to depart a little early.

But we were now airborne and headed to our next destination without any issues, hopefully meaning we wouldn't need to land back at The Hub.

As we settled into the long flight, a crew member needed to go to the bathroom, like most of us often need to when flying long hours.

The airplane was dark to allow people in the back to sleep, and Frabricio, an aviator not usually assigned to our crew, went to the back armed with his flashlight to make sure he was not going to be creating a bigger problem at the toilet (which would have been caused by poor aim).

He failed. As he got his clothes in order, he accidently dropped his flashlight into the toilet. We were the last plane selected to leave The Hub because we were taking the unit commander home, which plainly meant the commander was aboard. Fabricio, staring down at the toilet drain, the darkness swallowing the light of his flashlight like a black hole, knew he was going to have tell us, the crew upfront, and then his squadron commander.

He travelled back to the front of the airplane, climbed up the staircase to get to the flight deck and grabbed a headset to speak to us. Fabricio looked visibly concerned.

FABRICIO: Hey guys… So I dropped my flashlight down the toilet drain.

CARTER: Come again?

FABRICIO: I dropped… my flashlight… down the toilet.

BRICKEY: At the risk of sounding insensitive, first of all, why are you telling us? And second of all, how the hell did you do that?

CARTER: McKay! Please tell me this isn't going to break us! Please!

Carter had noticed on the mission computers that the airplane was still 20 minutes from the border point. Which meant that this could potentially force the crew to return to The Hub. Members of the crew had already been left on the plane that was left behind in Clearwater with the broken engine back when the deployment kicked off, and Carter knew this was a sign their luck hadn't necessarily changed.

Daniel Carter's girlfriend had planned to meet him at the recovery but couldn't stay longer than the night of the return before having to return to her home where she had to work at her neuroscience lab. They were a long distance relationship, and the redeployment date was slated for a weekday, making it slightly more complicated. The long-distance relationship was the result of her rising success in the field of neuroscience, and like all hardworking scientists, she had responsibilities during the work-week that couldn't be delayed indefinitely due to a flashlight breaking a military combat aircraft. Carter was being selfish, but he also knew no one would want to return to The Hub because of someone dropping their flashlight down a toilet.

As a silver lining, Hammond, Sheppard and Telford were on the other plane headed home ahead of Carter's plane, and their wives and girlfriends were going to going to be meeting them on time. But at the time, that was not Carter's concern.

McKay started to look into the aircraft diagrams to determine how bad the damage could get. If the flashlight got caught in a pipe bend, could a buildup of pressure cause a pipe to burst? He flipped through books and iPad pages of the digital flight diagrams, but engineers and book publishers had not anticipated anyone caring about the toilet piping. He simply could not conclude what the risks were. Regardless, McKay diligently took a solid 20 minutes to come to that conclusion to make sure Carter got the most accurate information.

When he did finally reveal this to Carter, Carter was able to reveal the good news:

CARTER: We're outside of Afghanistan. Combat Exit Checklist!

The crew cheered and passed the word along to the passengers in the back, which inevitably got the commander to come upstairs to celebrate. He was in for a surprise. Fabricio told him about the current concern. But at least now, it would force the airplane to remain in Syldavia, waiting for a repair team, not Afghanistan. But that would make Carter arrive late for his reunion. In an uncommon display of authority over a fellow combat aviator and equal ranking officer, Carter spoke up with instructions to Fabricio.

CARTER: Fabricio, you're going to have to go down there and try to grab that flashlight back.

Everyone stayed silent, which meant they were agreeing with Carter.

So Fabricio opened the small compartment beneath the bunk bed and another in the galley to grab trash bags. Lots of trash bags. He took off his uniform blouse and then his t-shirt before walking, chin high, bare-chested, to the back of the airplane, right hand and arm wrapped in several trash bags. Fabricio, the mighty weightlifter who could bench-press an airplane, was sent to the back of one by a skinny boy, to stick his hand in the back hole of that airplane. Literally.

When Fabricio got to the aft of the aircraft, word had spread across the airplane of Fabricio's new instructions and a crowd started to gather around the toilet to observe the proud moment. With the courage used to face down America's enemies, the military officer thrust his hand down the toilet hold. And it stopped!

The strong forearmed Fabricio could not get his arm past the toilet hole because of the musculature in his forearm. His hand swung wildly into nothingness inside the toilet, it not being long enough to reach the bottom of the toilet. He pulled out his arm and again tried to shove it back down but no avail. The large arm was simply too big to fit. This was a problem he never thought he'd have to deal with.

Fabricio was a strong officer in our unit. And he wasn't going to let this incident change his reputation as being an officer with a strong work ethic and a dedication to the mission that was difficult to parallel. There had been whispers from different offices in the unit wanting to hire him when he returned from the warzone. This wasn't going to change that. And he was a loyal friend. Fabricio knew Carter had a very serious reason to be back home on time and didn't want to let his friend down. He was willing to stick his arm down a toilet, after all, and tell the commander about it.

So he asked for a magnetic stick from one of the maintainers that had stood around to watch. The magnetic sticks were used by maintenance workers to grab metallic objects in hard to reach spots and that's exactly what it was going to be used for again.

The metal arm was stuck down the hole, but nothing could be grab. It slushed around, but as far as Fabricio could tell, there was potentially a small bend that made the metal arm useless. An actual human arm with joints was going to have to go down there. Fabricio knew it wasn't going to be his arm, so he looked at the skinniest maintainer he could find, and asked him to do it for him. The poor maintainer didn't know how to refuse a Captain (a superior officer). So he also stuck his arm inside the layers of trash bags and stuck his hand down the hole. For nothing.

After a few more minutes of suspense, it was concluded that nothing could be done about the flashlight. Fabricio returned to the flight deck with the news. McKay had continued his research and came up with the name of a maintainer who might know about this; he asked one of the loadmasters to find the maintainer and have him report to the flight deck.

The maintainer had been sleeping throughout the ordeal and therefore was unaware of the issue until he got on the deck. But once this young man was made aware of the situation, he simply explained the piping inside the aircraft lavatory. The entire system simply fed into a tank and the tank had a valve that was used to service it. There were no bends or narrow endings that could potentially be damaged from the flashlight, so it really should be of no concern.

After a momentary pause of incredulity, everyone let out a sigh of relief. We nodded and focused our attention back outside the windows. The aircraft was now flying over another country and in addition to the relief that the aircraft would not be landing in Afghanistan tonight, everyone was glad to know that we would be back in the United States on time.

We continued to drone on for hours. Our usual flight for the past few months had, at most, lasted 2 hours, with the occasional exception lasting longer than that. But this flight was 8.5 hours long. We flew over some allied states, and other friendly-but-not-too-friendly nations. While the flights over to Afghanistan were useful in getting to know each other and the crew, part of this flight was also used for this purpose, since a portion of the crew had been mixed up. A lot of the conversations also centered on what was coming next.

The crew was headed into a whole "new" world. Tonight would be the first night we could walk outside our rooms without a military uniform on, nor a gun strapped to our leg. It was the first night we would have our *own* rooms. Alcohol would be allowed, we could interact with non-military people, food would be served on a plate with silverware (discounting our pit stops at Klein Deutschland) and we could go to sleep without having to worry about whether or not a rocket attack would take place (no body armor vests would be lying next to our beds). For us the war would end tonight – we would simply be travelling tourists now. And like any good traveler, we had to decide what we would do. Like any good military planner, we decided we would improvise as we went.

This was not because we didn't have our own dreams of what we wanted to do when free from Afghanistan, but simply because we didn't know anything about the city we were heading into. Much of these vocal daydreams were focused on what we were looking forward to do when we returned home, but not necessarily to the Occident. So as the plane continued to fly forward, we talked and learned more about each other through these means. Some were excited to see their kids (obviously) and therefore we learned more about them through that – they would want to go hiking or skiing with the family, others would go to the beach, others had plans to enter isolation and go hunting alone in the massive national parks of the American west.

Seven countries overflown later, we finally arrived at our destination airport. Contrary to Afghanistan, where everyone we interacted with spoke English (we only interacted with fellow invaders, most of the time), we were now back in a world of English being a second language. This was made ever-more apparent with the controllers we spoke to, and their broken accent. Fortunately, the excitement had us overlook much of the incomprehensible words over the radio. So as the evening just started to begin, the main gear touched down at our new home for the night.

We parked on a separate ramp away from the other civilian traffic of the international airport, which was busy being the main international airport for the nation's capital. After shutting down our engines, a service vehicle arrived to service our lavatory. The hose was plugged and the pump was started. It was then shutdown, without any sort of damage or problems. As far as the crew was concerned, the flashlight was now off causing havoc somewhere else. But our plane was safe and the maintenance group that was with us sealed the plane for the night. We then all headed for a shuttle-bus assigned to the private air terminal we were parked at.

The section of the airport away from commercial traffic was an added bonus; we stayed out of the way of the airlines, but that meant we parked in the private air terminal, where private jets were parked for rich people that expected plush services. Our dirty, camouflaged uniforms did not match the usual attire the terminal attendants expected, and our glee was probably more visible than what most millionaires liked to reveal.

We certainly did not care!

Glass tables were surrounded by faux leather seats. There were bottles of water written with something other than Arabic or Pashtu, and there were glasses next to the bottles! There wasn't any dust (except for the quantities we brought in) and the high ceilings of the business-like lounge had two ornate chandeliers hanging from them. This was a great place to wait for the last straggling maintainers to get their overnight bags from the plane and join us for the ride to our hotel.

Twenty minutes later, after the shuttle-bus ride, we arrived at our luxury hotel and proceeded to check in. Again, thanks to the other plane that departed and arrived the day prior, the hotel staff was ready to welcome our bus load of people ready to create just as much chaos. Some of us decided to wait our turn to check in by going directly to the hotel bar at the center of the hotel, the nearest glass ceiling being eighteen stories above our heads in the open aired atrium. A waterfall and tempered-glass covered pool that visitors could walk over on their way to sit at the bar added to the luxurious decorations. This bar was where the first alcoholic beverages for the majority of the crew were purchased. Most also knew that the people who were the first to buy the alcohol were going to be the first to go to bed. We suffered from months of alcohol-deprivation, which meant everyone's tolerance to the legal psychoactive depressant was greatly diminished. So when a young 22 year old thought he could pound shot after shot when at home, tonight that 22 year old veteran would probably top out at 2.

After checking into our rooms and cleaning up, everyone broke off into their own group. The three primary officers that flew the plane over were together, while the loadmasters and some maintainers stuck with each other. The commander and few other unit supervisors went where high ranking people go.

One of these groups met 30 minutes after checking-in and started to walk through the lively city. Resembling most European capitals, the three walked along the sidewalks by cafes and bistros. There were so many restaurants to choose from, it was difficult to decide where to eat – after months of dreaming of this, we had no idea what to do. But we wouldn't be home for another 3 nights, and tonight was the first of those three. So we had to make some decisions.

## Phase II – Reintroduction

The group settled on a large, rowdy, hearty rustic restaurant that was founded in 1879. There, we ordered a hearty meal and laughed along to different stories of our experiences in combat. The discussions also focused on the future, still. What would be the first things accomplished when we set foot back in the United States? Drive my car, said one. Grocery shopping, said another. Eat sushi, go skiing – the ideas kept flowing as the crew continued to eat various courses at the loud restaurant. The lively atmosphere was perfect to drown away the excitement of the crew, with other patrons sharing in their own moments of happiness and enthusiasm. The team didn't sound any louder than the surrounding diners. The war, for now, was over. It was time to celebrate the little accomplishments of the past months.

In the back of our heads, the crew knew that tonight we would be sleeping in a room alone for the first time in several months. Between tents and containers, each of us would have access to a five-star hotel room complete with a large bathtub, potable water, reliable plumbing and electricity, and manual control over a thermostat. Even though we didn't think about it, there was a relief being out of the warzone that didn't necessarily relate to the dangers of mortar attacks, Afghan-turncoats or even aircraft accidents. It was just returning to a more 'normal' lifestyle.

After dinner, we continued to walk through the Parisian looking streets of Sbrodj and through small pedestrian streets found a cozy hookah bar. Within the bar, we shared a hookah, surrounded by a cosmopolitan crowd – non-uniformed people, young men and women mingling... Scenes only witnessed in movies for the past few months. It felt like a movie.

The simple things in life that are often forgotten were slowly reintroduced to us. Asking for a table to sit at, ordering an alcoholic drink, handing a credit card to pay for a meal, using cash to pay for a drink... Even bar hoping. We had reliable internet in our hotel room, personal showers and bathrooms, and no dust in the air. We could wear civilian clothes, and the most noticeable different from our months of wearing our uniforms were the shoes. In other words, we weren't wearing boots. The only time we wore other footwear than boots was when we'd wear our workout shoes. But private, normal, everyday shoes still feel differently and the sensation of switching from boots to civilian shoes after months can be compared to walking in normal shoes after a day of wearing ski-boots.

The next morning, the crew rejoined at the checkout counter and prepared to fly to another part of Europe: Scotland.

This flight also brought some differences from what we had experienced as 'combat' aviators for the past few months.

The sky is a directionless element. There are no convenient reference points like a lake on the ground or a mountain through which to find our direction. There's the Sun on a sunny day, or the stars during a clear night, but what to do during a cloudy day?

To help us find our way, the world has waypoints. These are different than the killbox keypads of Afghanistan. Instead, they are very precise imaginary points in the sky that all have a five letter identifier. Controllers and other pilots all know where that point is: it's the same on all the maps and it's free for everyone. These words are what was different from our combat experiences. While the radios come into our ears crystal clear, we also get instructions to fly to specific points that were different than our 'free-wheeling' abilities in Afghanistan.

Looking at the maps of our route, Carter chuckled at some of the waypoints he was marking on the charts and into the flight computers: VADER, KONAR, DORKI, ACORN, APPLE, or even MAGEC, BIDAC, ORTEZ, RILON. For American references, the arrival points leading to Andrews AFB (the base home to Air Force One) read as: HONNR, BRVRY, COURG, PLDGE, WEWIL, NEVVR, FORGT, SEPII. Meanwhile, SMOG is a point near Los Angeles, WAVEY, a point over the Atlantic leading to and from New York, and in Portland, their NBA basketball team, the Trail Blazers, are immortalized in the points TRAYL BLAZR. Kansas went the SPICY BARBQ RIBBS route instead.

Carter knew that those names were meaningful to the regions we would be flying over – cultural icons encoded into the uniform and standardized world of aviation. Gone were the coded military words to explain an airdrop or airstrike, instead replaced with accented controllers telling us to fly to a point that was named after an important landmark or event of their home countries. We had spent months in another country but rarely were exposed to the true cultural differences of the nation we invaded. But now, we were faced with a jumble of words that meant nothing to us but plenty to others.

After several hours of flight, overflying the entire span of Europe, our aircraft glided low over tumultuous waters of the Firth of Clyde, crashing waves rising up almost as if trying to reach us. We then turned around back towards the Scottish coast, overflying small islands of green pastures under the grey clouds and settled down at a small airport on the edge of the water.

Tonight we were going to be staying in a small Scottish town, a marked difference from the cosmopolitan capital of the day before.

This however, did not mean that we were still not excited to readapt to life outside of the warzone. It had rained for most of the day in Scotland, which meant the roads were still shiny and some puddles pooled in various areas of the paved streets. Such a sight had been rare for the last few months of being in either snowy mountains or dry deserts – not to mention that almost all paths (except for the flight line) were made of gravel, dirt and crushed stone. Finding a paved road to walk on was nearly impossible while deployed. The smells from grass and cool damp streets were a sensation everyone was smiling and laughing about on our way to a new restaurant to taste: a restaurant in an old, abandoned church.

While this was our second night out, it's important to remember that for most of the day, we were still in our military uniforms all day, flying in our military aircraft and conducting normal flight procedures all day – the same as we had for the past several months. Re-adapting took some time and habits can be difficult to change.

Consequently, eating in the old, converted church was another 'unique' experience that changed our perception of what normal life was. We were being served food *at* our table, we had to *use* our credit cards, our drinks were glasses and our food was served on a plate. Food was different from what we had eaten at the DFACs, with the few exceptions when we would travel to other sites across the theater of operations. Our maintenance troops were the ones certainly experiencing a different source of food for the first time since in a long time.

After dinner, we went out to a casual speak-easy, because we could. Just like the night before, we could walk around in our civilian clothes (which were the same we had worn on our way to Afghanistan and kept in storage, unused during our long months at The Hub). We were able to mingle with Anglophones (although we were still, for the most part, obvious foreigners, since the Scots would tell us we had accents even though we thought we spoke perfect English). We migrated to another bar near the hotel, stopping by an Indian window-restaurant beforehand. At the next bar, some of the crews dressed in their Afghan best; suits they had purchased from local tailors while in Afghanistan. By then, a state of inebriation officially impossible to attain at The Hub was reached by some of the military members returning from the warzone.

Which meant it was time to end the party for the night and head back to the hotel.

After interacting with a few locals though, we were slowly changing our tune from only thinking about our 'heroic, warfighting' exploits to 'everyday' issues, such as the cost of living in a town such as the one we were visiting. While the night before returning deployers would reminisce about their flights into hostile LZs or airdrops to SOF in the dead of night with radios crackling with action, our second night out was much more subdued, and instead focused on laughs and learning more about those who were kind enough to host us in their small corner of the world.

The next day, it was time to once again fly over the ocean.

We departed Scotland in the morning under an overcast sky, and unlike our flight over the ocean toward the warzone, this return mission would take place in broad-daylight. After piercing the tops of the clouds and seeing sunlight for the first time of the day, we noticed two things: the Scots would not be seeing the Sun today, given the extent of the single cloud layer, and why it wasn't moving. The winds were calm, which was abnormal. A massive thunderstorm in the north Atlantic, while no longer a factor for the crew, had left behind calm air all along our route to the North American mainland – this calm air meant that no wind was prevalent enough to move the thick cloud overcasting the British and Irish islands.

As with the first crossing, the crew continued to monitor all their systems and calculate their position whenever some of these failed. But with the calm winds, this made the math much easier.

Hours later, we slipped below the clouds were within sight of our destination. The name of the checklist called for was an exciting moment – it signified the end of a long, long flight over the vast ocean and one step closer to home. It was the last airborne checklist we would perform for this flight, which meant we were moments away from landing.

P: Ok, before landing checks.
E: Seat belt and shoulder harness?
P: Fastened and unlocked, pilot.
CP: Co.
E: Nav's talking on the radio but I've got a thumbs up. Engineer. Altimeters?
P: Set, 3022, pilot.
N: Set, 3022, SCNS set, Nav.
E: Radar altimeters?
P: Set, 200, pilot.
N: Set 200, Nav.
E: ADI mode and NAV SELECT panels?
P: I've got VOR/ILS 1, on my side, Pilot.
CP: Set, SCNS, Co.
E: Flaps?
CP: 50%.
E: Standing by on the gear.

A few moments later, we were in position to drop the gear, getting one step close to landing.

P: Speed checks, gear down.
CP: Speed does check, gear coming down.
E: Landing gear?
P: Down, indicators checked, pilot.
CP: Co.

E: Engineer. Landing light panel?
CP: Set, Co.
E: Hydraulic panel?
CP: Set,
E: Before Landing checks are complete.

Touchdown. We were back on the same continent as our loved ones.

We landed in Canada, onto a snowy runway and parked between two piles of shoveled snow. We were again welcomed to the same hotel we had lodged in for our trip to Afghanistan – the first 'familiar' site we returned to (none of the previous pit-stops had been to places we had been to before).

The squadron commander, familiar with the town after many stops to this point in Newfoundland over many deployments and years in the global mobility world, had organized a small party for the returning deployers. He had contacted the owner of a small Irish pub to get everyone returning from the deployment a discounted meal at the restaurant. Since the commander said that was an option, there really wasn't anywhere else the crew would go for dinner – if your squadron commander suggested a place to eat, it probably was wise to go even if it wasn't mandatory, although that didn't stop some from being frustrated.

Some still went around to other restaurants before the start of the 'mandatory' dinner, exploring different sites. A restaurant on the bay offered a glimpse to a vast ocean – a vacation like opportunity.

But the flight over the ocean was tiring and for the most part, many of us just walked around the quiet town before dinner was to start.

Our last night away from home was coming to an end. Some stayed at the 'Commander's pub' to continue drinking and celebrating while others went to other establishments in the town. Regardless, the return trip had been almost like an independent voyage between being home and being in the warzone. It was an emblematic opportunity, similar to our outgoing trip, for old friends from different crews to reunite and new friendships to form.

## Phase III – Getting Home-itis

This was it. The day to get home. The morning of our last flight outside of the United States. Nothing could stop us! We're aviators and airmen in the United States Air Force! "The greatest Air Force in mankind's history," as was often repeated to motivate the troops.

That great power was worthless in the face of some rain and wind. Weather reports were showing us we had a fight ahead of us for the rest of the day. Headwinds would be blowing against us the entire way home. These winds came from the Pacific Coast and gladly travelled all the way to the edge of this airport on the Atlantic coast. Which meant that to be able to take-off within the aircraft's limits, we had to launch off the small runway available at the airport. This small runway meant that we had less distance to use on take-off, which meant we had to be light. The opposite of what we were, given all the fuel needed to go home.

The front end crew poured over their binders, laptops and diagrams to determine how they were going to take-off. The wet runway meant that it was slippery. The short distance meant we had to be light. The long distance home meant we had to be heavy (filled with fuel). A pit stop was planned for us at an Air Force Base in Ohio, but that was still approximately 8 hours away given the wind forecast. We calculated everything to maximize our potential to take-off. The alternative sometimes mentioned was to perhaps wait for better conditions.

After a cursory glance, it was decided that was a terrible option, since it meant not going home as quickly as possible. We certainly couldn't allow that to be the plan. The U.S. military always got the job done. On time, on target. Especially when it meant coming home from war. Looking over the aircraft's performance charts, the weather and fuel requirements, we decided that the weather forecast was inaccurate (ignoring the conditions outside as well) and pushed to take-off from the short runway even if it meant changing some procedures. We'd get a solid idea of what the weather would be like when we would get airborne and hopefully above the rainy clouds.

Because the runway we chose to use was so short, we would be using maximum effort take-off procedures; unheard of since we left the tiny Landing Zones of Afghanistan's battlefields (which really was less than 3 days ago). Perhaps this was more unexpected than unheard of. The crew, skills honed by months of combat aviation and complex assault take-offs, was more than ready and capable, though.

The engines were pushed to the limit. The usual briefings were given to the crew, and the plane was lurched forward as it started its race down the shortest runway the airport had to offer. The other runways were longer but pointing in the wrong direction relative to the strong winds that were prevailing this day. The aircraft performance limits were clear: taking off with a massive crosswind like that was just too dangerous. Which meant that the aircraft performance on the long runways would have been more dangerous than using the shorter strip.

Rushing down the cold runway flanked by snowbanks, the C-130 we had come to rely on over the course of this multi-month deployment with countless (we actually counted, it's a little over 200) combat sorties and missions, continued to perform superbly and hopped off the ground, racing upwards into the cloudy skies as planned.

Only a few hundred feet up, the windows on the airplane served no more purpose, the clouds around us obscuring everything. Fighting the turbulence, the biggest concern the crew was observing were the six digits on the computer screen. The last time there was such a concern with those numbers had been months before when we had crossed the ocean: 270/110. The winds were blowing 110 knots, directly against our flight path (heading 270). That meant that our plan to fly 300 knots of true airspeed (KTAS) actually meant were going to be traveling 190 knots over the ground. This was a significant fight to face.

This fight was going to cost us. The price was gasoline. Our engines were performing for us to fly 300 KTAS and burn the fuel accordingly. We had planned to be flying faster than 190 knots over the ground. This meant that we had expected to travel a certain distance after burning a specific portion of fuel – and this was simply not going to happen. According to Carter, we were going to fall very short of our objective.

After calculating three times the fuel consumption and consulting a variety of charts and working his mathematical magic, we heard Carter get on the radio with the other airplane further behind.

<u>GRASP 01:</u> GRASP 02, GRASP 01.

GRASP 02: Go ahead, GRASP 01.

GRASP 01: Carl, I need you to check me. I've been going over the numbers here on the 4116 using 970 TIT with our current FF, uh, but I need a double check on the arithmetic.

GRASP 02: Yeah, you can go ahead.

Carter read off the numbers he was getting and his calculations to the other aircraft for it to verify on their end the concerns that were growing.

GRASP 02: Your numbers are good.

GRASP 01: Well that's not good.

Carter's statement was certainly an understatement. But it was fortunately still early in the flight. Flying thousands of feet over Canada, the crews agreed to continue their flight forward and re-verify our calculations in one hour. The hope was simply that the weather and winds would improve, negating the concern that had started to build up soon after the euphoric take-off from Newfoundland. But for now, the crews were being told that they wouldn't have enough fuel to reach their destination.

An hour went by and the winds and weather did not improve. The winds were still just as hellish and we could not see further past the aircraft's bulbous nose as we continued to fly through the grey clouds. The radar hiding behind that nose was telling us not to get our hopes up regarding the weather. Both crews on the two planes started to talk to each other, again. We were all facing the same problem and we had to start working for a solution. One of the airplanes had a small mechanical issue that wasn't a huge issue for the flight, but the regulation was clear: "the failed system shall be repaired at next repair capable facility."

This was not a concern when we made it across our different European stops. The airports were not repair capable facilities for our combat airlift aircraft. But as we saw our fuel reserves lowering without much ground gained, we started to discuss different airports we could land at to get more fuel. A military airbase was always favored – fuel would be cheaper and prior permissions to land were easier to acquire or could be avoided altogether. But most of the airbases we were looking at were also "repair capable facilities". If we landed at some of the fields we were looking at, it would doom one of the airplanes and it would not be going home today. Or for the next few days, since the plane would need time to get repaired. It would be back in the United States but still so far from its home airport. The crew could request a waiver, but we didn't want to bet on it.

The second challenge was the Customs piece. We were technically arriving from foreign countries, which meant we would have to in-process through U.S. Customs and Border Protection before moving on. That process had to be conducted at the first port of entry into the United States, which simply means the first airport we touch-down at. That airport would be our most critical stop, which meant we had to weigh our maintenance concerns, our fuel worries and our customs welcome service. International airports had Customs services, but regional ones did not, and most military airfields did not. Even if we did land at a Customs-capable airport, the time it would take for them to come process us was critical: if they took too long, we would be out of our legal duty-day to take-off and make it back to our home within our legal flight time.

The strongest contender was an Air National Guard facility in New York state that was co-located with an international airport. This stop was not optimal because it was a "repair capable facility." But it had the customs services that could be called upon quickly, and the fuel. It was close enough to home that we could reliably make it from there to home without too many more concerns. But we still searched and discussed alternate sites. Another airport had no maintenance but customs was not ready to welcome us. Other airports like Boston International Airport or similar sized ports were simply too big and important to deal with us. We could not just go bug them because we wanted a few drops of fuel to get home.

Looking at our charts, we determined we were going to be crossing into American airspace in 10 minutes, and the fuel situation was not improving. Nor were the winds. But that was when an epiphany finally hit both airplanes in flight.

Four and half hours after our take-off, a small airport in Maine would be our saving grace. The longer-than-two-miles-long runway was the oasis we needed. It had Customs (which was in the habit of processing returning military troops from the wars), it had fuel (it was an Air Force KC-135 air *refueling* tanker base) and it was not a maintenance capable airfield (it was a KC-135, not C-130, base). And it was only 30 minutes away.

Immediately, we got on the radios to start working our diversion plan. Some of us called Air Traffic Controllers to tell them where we suddenly wanted to go, and others of us got on the horn to talk to our new island of hope. They were ecstatic to hear that two cargo planes filled with military troops were going to be barging onto their runway soon and be needing fuel and customs quickly to then depart as suddenly as we arrived. That much was not clear in their tone of voice. But they were helpful and started their procedures to see that our needs were met.

Controllers helped us and got us pointing in the proper direction and within 30 minutes, we arrived at our destination.

This portion of the flight should have taken 4 hours per the plan we had originally hatched with the weather forecast we had used. As we agreed, it was wrong. We had assumed the winds were forecasted to be worse than reality. But the truth was the winds were far stronger than the forecast. So we ended up flying a 5 hour 10 minute flight. This meant that the next 8 hour 10 minute planned flight was actually probably going to take 10 hours 20 minutes. We consequently got on the phone with Sarah (the Unit Deployment Manager at Clearwater) to notify her of the change of times and our new ETA. Everyone else that wasn't part of the flight crew was already calling their friends and families to tell them that they were going to be late. No one cared. The most asked question was: "but you're coming home, right?"

Yes. We were finally on our way home. With the customs officials gone, we fueled the planes up to their maximum capacity with fuel. The runway was long enough to welcome a Space Shuttle, so even if we were very heavy with fuel, we would have enough concrete road to take-off from.

We shut the hatches and doors to the planes, started the engines and quickly returned back to the cloudy skies over the United States. Our first stop on American soil was as quick as it could be – no one cared whatsoever to be back in America except that it meant phones could finally be used without dealing with roaming or international fees. Other than that, the aim and view was still pointed west to our home airport.

As we flew westward, controllers gave us the final call we had been waiting for:

Center: GRASP 01, and GRASP 02, cleared direct Clearwater.
GRASP 01: GRASP 01. Cleared direct Clearwater.
GRASP 02: Zero-two: direct Clearwater, thank you.

The computers were updated and we pressed Enter. Instantly, the aircraft banked a little left and had a single, 1300 mile line counting down from us to our home.

## Phase IV – Home

Now that we were cleared direct, we started to focus on more important things: like *when* would we get home. Analyzing the winds and using some quick math, it was decided that we would arrive in about 10 hours. So we all quietly settled into our seats. The flight home was as simple as any airliner flight. We overflew the Great Lakes and used the radar to "see" Chicago (since clouds prevented us from seeing anything outside).

A few hours out, one of the crewmembers decided to call his wife from the airplane. Instead of using the satellite phone we often used in Afghanistan, he was going to use an HF phone patch. The technique consisted of using the aircraft radio to call a service station. That station would dial a phone number we gave them before connecting our radio call to the phone line. Using this system, we were able to call this crewmembers' wife's cellphone from 20+ thousand feet while she was driving through Clearwater to welcome us at the regional airport.

Mrs. Crewmember was of course excited to hear from her husband, and was informed about our progress. She would relay the details to Sarah and the rest of the cadre of officers and members helping in the recovery effort. We then terminated the HF phone patch and continued our two-ship journey to our final destination.

The last few hundred miles were spent cloud-surfing. The weather was now improving for us and we finally saw the Sun. It was setting and casting a gentle red glow over the nap of white and orange clouds a few inches below us. Our altitude was the same as the top of the clouds which meant we were like a snowplow in a fluffy world of cotton. The setting Sun seemed fitting for the end of our long journey – both over the past few months and simply over this 18-hour work day.

We finally reached the airspace controlled by Clearwater and started to talk to them. We didn't know if they were aware or not we were coming back from the "sand-box" (a term used by 'cool' military people referring to the Middle East) but they quickly made it known they didn't:

Approach: GRASPs 01, 02 proceed to the Clearwater 360 at 20 and hold.

GRASP 02: Um… GRASP 02, direct CLW 360 at 20 and hold.

N: I'm just going to use my expert opinion here and tell you guys that this is not taking us in the right direction.

We proceeded to fly directly to a point about twenty-five miles north of the airport where we would fly in a circular pattern to hold over that point as the Sun completely disappeared over the horizon and night set in. The reasons why we had to hold were never revealed, but fortunately, the holding only lasted a few minutes.

Approach: GRASP 02, you're cleared to set up for a five mile final for the visual to runway 27.

GRASP 02: Cleared for a visual 5 mile final to 27, GRASP 02.

Approach: GRASP 02, switch to tower, 23.5.

GRASP 02: GRASP 02, switching to tower.

Five miles out, we made a right turn to line up with the runway, and soon found our main landing gear kissing the concrete runway. We had finally landed. The runway fed directly to our parking spot, so we avoided hitting the brakes as we raced down the runway as quickly as we could. We could *see* our home base – its hangars and operations building.

After a few more left and right turns, we followed the marshaller's light stick in the night to line us up with our parking spot. Further behind the marshaller were dark crowds, all bundled in their jackets, huddling together against the bitter cold, and bouncing around in excitement. A kneel of the front wheel of our aircraft showed we had come to a complete and final stop. The crowds were freed from their invisible holding pen and started to approach the airplanes a little bit (but not too close, until we shut down the *loud* giant fans that were our propellers).

For the last time since we departed months ago, we shut down the aircraft's engines without expecting to restart them again for a few weeks.

P: Engine shutdown checklist.

CP: Nose-wheel and Parking Brake

P: Centered, set

E: Times are good, certified McKay, informing the crew that shutting down the engines would not damage them.

CP: All four, coming down.

The powerful propellers were shut down and let out a relieving, spooling down whine.

CP: Radar altimeters?

N: Off

P: Off

CP: Oxygen

Crew: Off.

P: Static left.

The propellers on the left were at a standstill, which was the signal that the rest of the checklist could continue provided the propellers on the right side of the aircraft were also done spinning.

CP: Static right. Exits?

P: Clear - Insert Chocks

L: Chocks in place

CP: Parking Brake

P: Released

CP: Hydraulic panel set and lights off. Engine shutdown checklist complete.

Out of the cockpit windows, we could see in the middle of the night families and vising dignitaries starting to walk closer towards the airplane. The governor was again there to welcome us home, and so were the highest levels of military officers within the state's Air National Guard. Behind the distinguished visitors were the more important visitors: friends and families that had waited for hours. Months, in fact. Months for us to get our act together, get the job done; months of hearing complaints about our living conditions, use the "but I'm at war" excuse on the phone to force our loved one to not get off the phone just yet. Hours for us to decide whether or not to divert to one airfield or to another. Months of life that happened without us, but with our memories constantly in the back of their minds, subtlety reminding everyone that someone was potentially in harm's way.

With everyone outside waiting for us to get out, McKay was rushed to shut off the airplane power-systems and with his final flip of the switch, the airplane went completely quiet. And dark. All the lights went out and the flight deck was only lit by the outside lights seeping through our cockpit windows. Normally, we would have a power cart to provide electricity for basic aircraft systems, but where we usually plug in the cart was blocked by the Governor and the line of officials that waited by the door to shake our hands.

The Crew Entrance Door was soon opened and like a firehose, everyone from the back exited the airplane quickly. One hand after another was shook and at the end of the line, we finally came face to face with our families and friends. As expected, people ran toward each other, hugged and kissed, kids were taken into arms and some tears were shed. Friends from the unit, from across the country had made it to welcome back all of us. Girlfriends and wives jumped into the open arms of their loved ones.

Sheppard was finally reunited with his wife, having spent the majority of his marriage overseas, taking part in combat operations. He was ready to spend more time with his wife and begin and share his new life with her. Ford and McKay were ecstatic to finally start cruising down the state highways with their motorcycles. Telford hugged his fiancé and looked forward to finally tying the knot. Fabricio was going to be joining a shadow world of aviators in a program whose existence is not even acknowledged, the toilet incident far behind him. And Hammond was thankful he had finally completed his Air

Force Short Tour service, preventing him from having to leave his wife for a full year overseas. He instead kissed her hello, both of them knowing they were about to start planning their move abroad to live in Germany. Together.

We were finally home. The war, for us, for now, was over. We didn't dare think about when we would have to return. Some of the crew would head back out the door in a few months again. Others were moving to different assignments. But none of that mattered. We now had to rebuild our lives.

And we had returned to a world where we were free, but invisible.

## The Disconnect

We were now home.

Coming back into work, we'd hear the occasional welcome backs, but otherwise, everyone went about their day. It was our responsibility to be re-admitted into this unit that had lived perfectly well without us for the past few months. We were just added bodies now. Soon, another group would be leaving for the warzone and we'd be taking over their responsibilities. Our multi-month vacation to Afghanistan was now over.

The reality about deploying with the military was that all military members in the unit had probably done it. And everyone had always done it better than you. No military officer, let alone aviator, that had to claw themselves to the ranks of the elite, would ever accept being second-best. For the past decade, or so, of war that the United States had been waging, from Africa to Afghanistan to Europe and Asian support operations, everyone had done it all and seen it all. The old adage of 'one team, one fight' was repeated to us for a reason: we could not do our job without others. We only rose to any challenge together.

Consequently, no one individual ever did anything special. Very few ever did anything incredible, because then that would mean there was a period that was less important, and therefore, you didn't know as much and your contribution was minimal. Less important. Some members from the squadron that had deployed before us didn't abide by the unspoken rule that you don't talk about your deployment. Instead, they would embellish it: WAR! Danger at every corner. Death staring you in the face. But very few respected those who did that. However, knowing about these people had prepared our crew well for our own expectations when we returned. We would not repeat the same errors.

Because of how regularly the members of the C-130 community deployed, adapting to the United States is in and of itself just as complex as adapting to life in another nation. Deployments were just as long as the time we had at home before returning overseas; it what was called a 1 to 1 deploy to dwell ratio (if you were deployed for 6 months, you would be home for 6 – if you were gone for 4 months, you'd be home for 4 before heading back out for another 4 months). Such was the scheduling cost of the continuous requirements placed on the military to sustain the war-effort. The 'home' time, though, was still peppered with various smaller missions (called TDYs – temporary duty) that could last from a few days to a few weeks.

Returning to the United States, new challenges appeared that for some may seem benign. We had to remember to buy groceries at the supermarket, since we had kitchens once again. There was no free cafeteria to access anymore – budgeting had to be adapted. We had to focus on and pass physical fitness tests while also conducting maintenance on our cars (which had been sitting in a garage for multiple months). People's failure to understand what our minds were going through were met with our own failure to readapt quickly to life in the United States. Members of both parties in relationships were sometimes to blame for frustrations – we would judge overweight Americans we saw in supermarkets, becoming the judgmental and extreme controlling monsters we tried to stop abroad. Not to mention, we were part of someone else's life again – not our roommates from the desert but our girlfriends or boyfriends, wives, husbands, children and other family members. They wanted to be a part of our life and they wanted us to be a part of theirs. But being part of our lives would prove to be difficult.

While we couldn't talk about the combat deployment at the office with fellow military members, we certainly could never speak about it to our civilian friends. It was tough to get anyone interested at that point about the war. To remind people why we were asked to do these things we do. That while some may have forgotten, we still are sent over to fight by officials we, as a nation, elected to office. What for? Sometimes, we had our own doubts to the answers: to help people get out of a bad situation – humanitarian responsibilities... Or perhaps to destroy extremism. How? We were not sure either. After, don't we have very real problems in the US that need more focus than just pretending terrorism is the biggest threat to the US? These ethereal questions that were tough to answer as a society meant that we simply didn't even bother asking them. Instead, we just accepted the random 'thank you for your service' comments and knew that being in the military meant that we'd go somewhere, for some time, to do something.

What we were sure about is that to describe the actions we had to take when flying the airplane would take too long for anyone to stay interested in the story. It took some time for us to become proficient at flying, so it had to take a little longer than a 2 minute snippet between two stories to explain the intricacies of combat situations.

If a loved one cared to hear about the deployment, they also could not understand. Not necessarily because they didn't care, but simply because of the complexity of the operations. How to explain the joke that Carter had said the destination name instead of the Killbox-Keypad to Tetra? That the static line retriever instead of the pickle button was hit? How to explain that the plane was so heavy we almost couldn't get out of an LZ? Context is everything (some small bush-planes would laugh at what we called a min-sized landing zone). Or, we could talk about how we lived in a container with another man. Showers clogged, an aircraft engine caught fire more than once, we were stuck in a bunker because of a bombing, but what would that serve? We would make our loved ones worry, and we still wouldn't feel better about what *we* accomplished.

That was something that we had to find within ourselves, and for the younger members of the crew, it did take some time. Time to get over the pride that sought attention and gratification. It took time for this praise-seeking-pride to change to an internal pride – one that allowed them to face other challenges in life with much more confidence.

Not to mention future deployments.

Finally, the moment we took-off out of Afghanistan, and the whole way home, we were no longer a crew. Some may have given up on being friends before the end of the deployment, but the final landing at home is the irrevocable sign that no one was part of the crew anymore. Everyone went back to their old lives. Very few crews got back together for a beer or a dinner to talk and catch up. This was true not only of our deployment, but many that have come before and after us. What everyone focused on doing is getting their careers back on course: what's next for the promotion? The next assignment? The deployment box is checked, proceed. This is reflected in military reports, where only 2 out of 9 lines of our performance reports mention the deployments.

For the Clearwater unit, as a hard reminder that life moves on, it was announced that our unit was closing, and our Commander was fired. Feelings became polarized: some loved him, others hated him. It was fair, or it was not. The unit was closing, so we all had to compete for a slot elsewhere now. Some of us would have our choice made easier and were selected to fly for Air Force Special Operations, but others did not follow that trajectory. The commander was not the only person to have a career change – Ford was discharged from the military after other revelations were brought to light about his behavior. His contributions to the war effort were still critical and should never be discounted. He did far more than most Americans ever would, particularly when it came down to sacrificing personal time to help others.

So the war, in our minds, was now paused. We returned to training, and we continued to prepare for the next time. But we never spoke about it. We instead perfected our flying and our techniques, keeping our experiences in the back of our minds to refine those skills. But rarely did we hear someone say "we did that during the deployment." It was an interesting way the community handled it, but we all respected each other's deployments without ever talking about them. We had earned the leather jackets (which we now wore), we all flew together at home, and we all had experienced various events that either are classified or just slightly different from others. We understood that, which was why trying to speak about our experiences sometimes looked like an individual was trying to 'one-up' you. If someone spoke too often, they became a "so there we were..." story teller. The person with the story.

Everyone had a story. No one wanted to hear it.

An ironic statement, given the creation this book. Only these sheets of paper and its readers will know of the events in Afghanistan we took part in, except for the involved crew-members.

## Final Note

While living abroad, we still stayed 'American' as much as we could. As mentioned before, our days off were often broken up into different TV watching periods. Two hours would be spent watching two 40 minute episodes of The Walking Dead and one episode of Friends, or three episodes of The Big Bang Theory, or two episodes of Entourage… The variety was colossal. By the end of the deployment, a single crewmember will have watched six seasons of The Big Bang Theory, read all the Harry Potter books and watched all of the movies, two seasons of Game of Thrones, four seasons of The Walking Dead, all Star Trek and Star Wars movies ever produced, two seasons of The League, and two seasons of Stargate Atlantis, along with a dozen different films. This seems like a lot, but it's been researched by the Nielsen rating agency that the average American, in fact, watches 40 hours of programing a week, which equates to an average of 5 h 45 min. of television a day. This was about what our crew would watch in a day. But we worked every other day, which meant that with all of our slouching and bed-surfing, our crew still did not manage to meet the American average. It was closer to three quarters of the average Americans consumed (although sometimes we had a little time after a work day to watch some TV at home. This would occur if we everything synched up perfectly and we could run early and the cargo would still be ready. The earliest we've ever been allowed to depart the port terminal was 2 hours and 20 minutes.)

The final tally of accomplishments the crew had completed after our deployment to the Middle East were calculated on the flight home, now with the certainty that there was no more cargo to be carried in Afghanistan.

Our crew carried about 1,500 passengers interspersed over several hundred flights. It could have been done with 18.8 C-130s, fully packed with passengers. Or just 2.4 Airbus A-380 aircraft (such as the ones Air France flies from Paris to New York every day!) Cargo wise, we airlifted 800 metric tons of cargo (about one million, seven hundred pounds). That was over hundreds of flights. About 5.6 Boeing 747s could have handled it… Overall, it was a very small contribution to the war-effort.

But that small contribution was added with two slightly more significant contributions.

None of those airliner-designed airplanes could perform the airdrops. Special Forces teams in mountainous outposts were resupplied from the air at night against hostile forces and conditions. That was the first contribution.

The second is that none of the airliners could land at some of the airfields the C-130 could. Helicopters can land anywhere, but they don't have the range or cargo capacity to reach the far flung outposts. While some airplanes bigger than the C-130 *can* also land on dirt LZs, the added weight of the larger aircraft often threaten the integrity of the LZ which meant they *shouldn't* land on dirt LZs (they could land and takeoff once, after which the runway would be shredded). And the larger airplanes are so massive that they are often overkill for the small FOBs American forces have. So that was the second contribution.

While this story recounted some of the adventures faced during this deployment, there are still several classified missions that can cause our veterans to pause and think about the sacrifices and dangers they faced. Missions and achievements that will remain secret often for security purposes. These missions can thankfully remain classified – we don't need to worry our loved ones more than they had already been during their ignorant watch back in the United States.

And so, these are most of the things we did in Afghanistan.

Again, none of us ever pretended to be hardcore, centurion or Spartan-like, SEAL-Team Six, Delta Force and Ranger regiment warriors. We had survived only about officially recorded 80 attacks (bombings or shootings etc.) compared to what some others faced.

The reality of who we were in this war effort: the not-warrior folks. Just the simple Americans who chose to get involved in something. Some did it for the moral obligations we feel. Others, for the money. Others, for the job. But the end result was the same, whatever our motivation: we were all a small piece of this effort that was driven by our nation. In exchange, we vowed to do our job perfectly and whenever needed.

We understood that we are not the warfighters in the sense that we kick down doors or kill terrorists. But we were highly capable assets. We were able to sit with members of those elite teams, and hear them say to us: 'we trust you'. If we did our part, then they could do theirs. If they trusted us to get them part of the way there, then in a sense, we succeeded.

Our abilities and successes proved to United States citizens that they could count on us. At any time, day or night, we were ready to go and rescue a hostage, or airdrop food supplies to a hurricane-devastated region, or fight wildfires threatening homes.

We were symbols of hope coming from the air. A rare sight in the modern $21^{st}$ century, where Americans learned to fear planes crashing into buildings, and the rest of the world came to fear bombs coming from the same domain. We never dropped bombs. Only supplies to those who needed it most.

## Stories in Photographs

To avoid repetitive stories, below are a few photographs taken during the course of the deployment to help visualize operations. After all, a picture is worth a thousand words. Some portions of certain images have censored for operational security purposes.

Our C-130 lands on an unpaved patch of flattened earth to extract an Army unit. This photograph was taken by a USAF member waiting on the ground for us with his team. Photo credit: MSgt Ben Bloker // USAF.

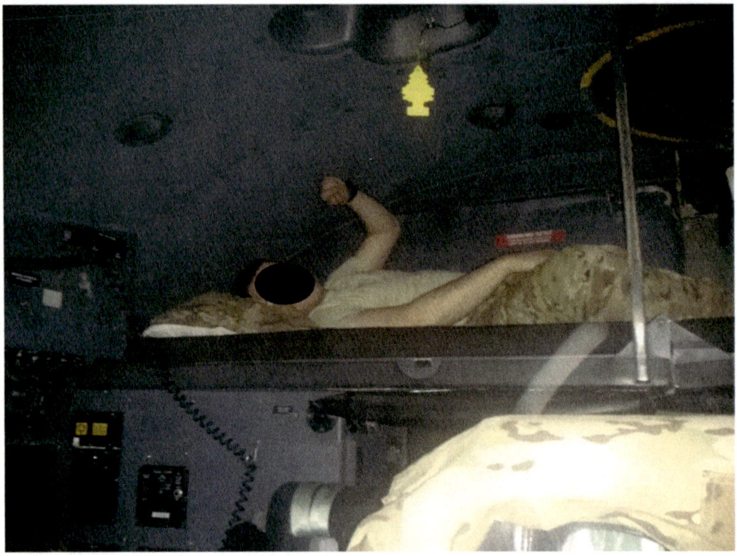

Sleeping in the C-130 after breaking down and being stuck, as mentioned in the chapter "Operational Challenges – A Night in the Airplane". Some FOBs were extremely small and had no additional rooms.

On long missions, the crew and maintenance team would hang their hammocks or set up make-shift 'beds' on the pallets to create a small private space to travel comfortably. A clever pilot would start to lightly yaw the plane left and right to rock the hammocks.

'Fabricio' searching for his flashlight in the aircraft lavatory. The search was conducted to make sure the aircraft didn't get damaged and ensure a timely return to the United States, as mentioned in the chapter "Return Home – Phase I – Escape".

Travelling over the North Atlantic and still about 450 miles from the nearest land, the GPS navigation system failed, as mentioned in the chapter "The Crossing – Second Leg – Over the Ocean". The crew's knowledge of their systems and airmanship helped counteract the 114 knot crosswind that could have easily pushed the plane off course without them ever knowing it.

The Transition Center was a base designed just for that: transitioning into the combat zone. Tents were used to house everything as we were hidden from view of the local population behind tall Hesco barriers. We weren't in any danger from enemy activity here, but we would soon be.

An abandoned Soviet aircraft with huge amounts of natural growth around it at the Transition Center, as mentioned in the chapter "The Crossing – Third Leg – Deployment Starts". American warplanes and military personnel operating here was a sight that, 10 years before Sept 11, 2001, would have been unthinkable.

The Transition Center was hosted at the airport of a former Soviet state, as mentioned in the chapter "The Crossing – Third Leg – Deployment Starts". Note the writing on the tarmac and civilian terminal sign written in Cyrillic.

The remains of a FOB evacuated and abandoned by US forces. Our crew was the last fixed-wing aircraft to ever be there, as mentioned in the chapter "The Closing".

After picking up troops from a dirt landing zone (a road), helicopters that were covering the troops moved out of our way so we could take off and fly away from the battle zone.

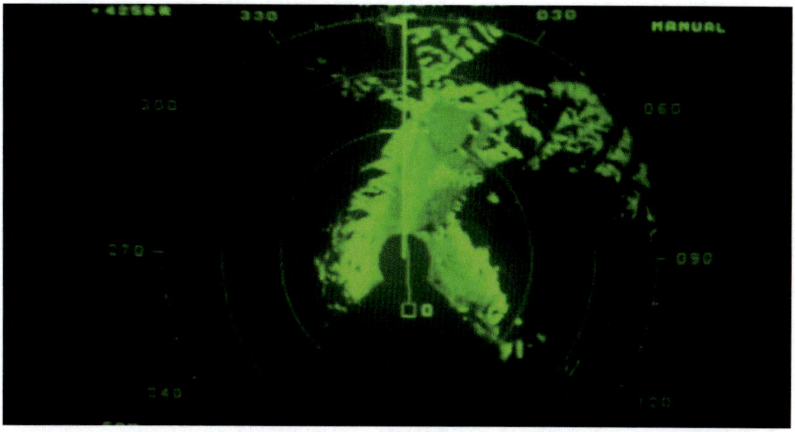


Radar image of a low level approach to a FOB, flying between two mountain ridgelines, well below the mountain peaks (which is why there is nothing but darkness beyond the edge of the canyon walls – radar energy could not 'see' anything beyond.)

The Hub at night. An island of light that could be seen miles away, the perimeter of the installation is clearly visible. These perimeter lights all looked outwards to detect any hostile forces approaching the base for a potential attack.

A view of the stars through our night vision goggles (NVGs). Can you see Orion's belt?

Some of the essential tools of a C-130 flight engineer: night vision goggles, checklists and TOLD (Take-Off and Landing Data) cards.

While Afghanistan is often viewed as a warzone with nothing to offer, that is not an accurate image of the country. Apart of the beautiful villages and oases scattered all across the nation, we also were able to view the remains of Bamiyan and its statues of Buddha carved into a mountain. These preceded Mount Rushmore by 1,400 years.

A firetruck responding to our emergency mentioned in the chapter "Operational Challenges – Happy Birthday".

The remains of a building 50 feet from our dorms at The Hub after a mortar attack against the base. This event took place one week before we returned to the United States. This photograph was taken at night so it's difficult to see, but looking closely, you can see some holes left behind by the pieces of shrapnel as well as the twisted pieces of metal caused by the heat and explosion.

Our crew was selected to transport a very important member of the government. Because of the hostile LZs and dangers associated with many areas of the AOR, high-profile personalities must give up their polished jets for combat airlift aircraft such as the C-130. But we still installed what we called a DV-pallet, which was a self-contained luxury suit that could rolled onto the airplane.

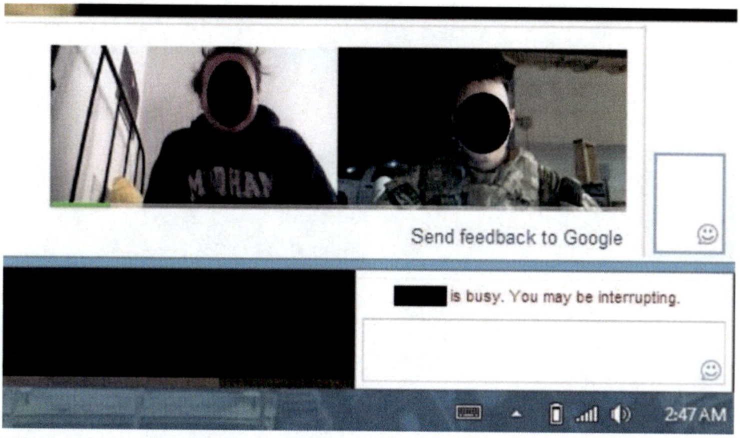


To keep relationships going, members of the crew would often go to outdoor public areas where good Wi-Fi could be found to video chat with their loved ones (there was Wi-Fi in the rooms, but leaving afforded some level of privacy). Sometimes a source of stress for those in the United States was seeing their loved ones wearing their body armor and guns while on the phone, like in this image.

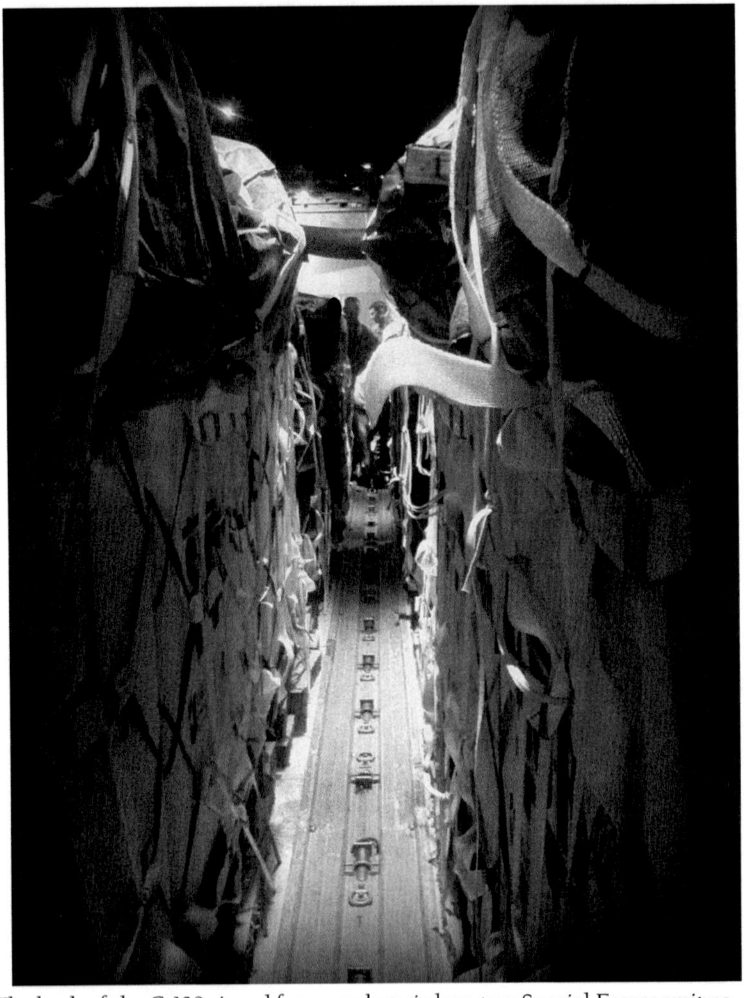

The back of the C-130 rigged for a combat airdrop to a Special Forces unit as mentioned in the chapter "Combat Airdrop". The black layer atop the white box were the packed parachutes.

While there were times when our crew could wait for a maintenance team, there were sometimes some realities brought forth by war that dictated the aircrew repair the plane themselves...quickly! This obviously to avoid a more difficult combat situation. Here, our crew rapidly developed makeshift tools to hand to another member that was inside the wing structure (see his boots on the wing?) while not far away and in the background, smoke billows from other '*activities*'. We were able to hastily repair the airplane and return to The Hub for more permanent repairs.

Being deployed didn't mean we had to ignore traditions. Here the crew wears Darth Vader masks on Halloween, the masks a gift from Carter's girlfriend.

A still frame taken from the filming of the number 2 engine torching as mentioned in the chapter "Operational Challenges – A Night in the Airplane".

An image of one of aircraft flight computers the moment after we were cleared direct to Clearwater on our return trip from the deployment. The Distance To Go (DTG) of 1339 nautical miles, seen in the left side of the screen, was all that was between us and the end of our multi-month deployment, as mentioned in the chapter "Return Home – Phase III – Getting Home-itis".

Cloud-surfing toward the end of our final flight home, as mentioned in the chapter "Return Home – Phase IV – Home".

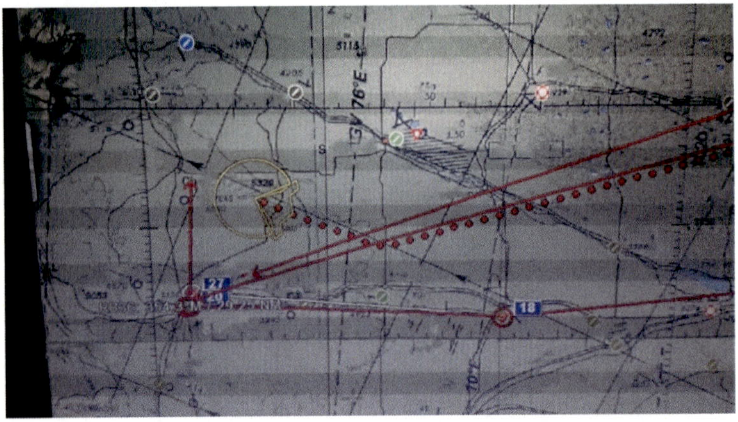

Just 30 nautical miles from being home after months away, Air Traffic Control tells us to divert to a point *away* from Clearwater, as mentioned in the chapter "Return Home – Phase IV – Home".

The End

## Acknowledgements

To you, the reader: thank you. The purpose of this book was, at first, therapeutic – to write down a piece of memory and history that we had the honor of taking part in. But it is you, the reader, who makes the difference. Your willingness to learn more about these events represents why our actions are not for nothing and helps us, members of the crew, feel less lonely. You are the reason why this is a book and not just a saved file on a computer.

But readers would not have been able to read this in the first place if it were not for those around me that helped me make this possible. My world is a better place thanks to all those who were willing to help me and lead me through life.

First, to my parents, Marie and Joseph. Very simply: I would not be who I am today if it weren't for you and the opportunities you allowed me to seize. Your willingness to support me in anything, even writing this book, has always been a rock to rest upon in the torrent of life.

To MJ, thank you for the endless hours of editing and rambling about the events that took place in the Middle East. Your patience with my random tangents and other 'so-there-we-were...' stories is inspiring and you helped keep me focus – thank you for your tireless encouragements and support.

To my baby brother; I'm sorry for not having been around very often, sometimes because of the stories contained within this book. Looking back on life, there are moments when we have to wonder if they were worth it, or if perhaps staying around to see others would be more valuable. You've taught me that exciting parts of life don't necessarily lead to better memories. Thank you for staying close and a part of my life – it makes it all much easier.

And finally, to her. Your steadfast support and love made these deployments more bearable, and I can only thank you for your patience. Thank you for your endurance, love and care.

So once again, to all of you: thank you!

Please don't let the simplicity of these words diminish my sincerity.

## Acronyms

Because the military loves MLAs (multi-letter acronyms) and could not exist without them, below is a list (with pronunciation) to remind you of the incredible confusion not naming things can cause:

| | | |
|---|---|---|
| ADO | A.D.O. | Assistant Director of Operations |
| AOR | A.O.R. | Area of Responsibility |
| AFCENT | af-cent | Air Force Central Command |
| AFE | A.F.E. | Aircrew Flight Equipment |
| AMC | A.M.C. | Air Mobility Command |
| AMD | A.M.D. | Air Mobility Division |
| AOR | A.O.R. | Area of Responsibility |
| APU | A.P.U. | Auxiliary Power Unit |
| ARA | A.R.A. | Airborne Radar Approach |
| ARMS | arms | Aviation Resource Management System |
| ATOC | A-tock | Air Terminal Operations Center |
| AWACS | A-wacks | Airborne Warning and Control System |
| AWADS | A-wads | All Weather Airdrop Delivery System |
| BX | B.X. | Base Exchange |
| CAOC | kay-awk | Combined Air Operations Center |
| CCAT | see-cat | Critical Care Team |
| CDDOC | C-D-doc | CENTCOM Deployment Distributions Operations Center |
| CDS | C.D.S. | Container Delivery System |
| CENTCOM | cent-com | Central Command |
| CFACC | see-fac | Combined Forces Air Component Commander |
| CFR | C.F.R. | Crash-Fire-Rescue |
| CJSOTF | C.J.sotif | Combined Joint Special Operations Task Force |
| CTII | C.T.2. | Combat Tracker 2 |
| DFAC | D-fac | Dining Facility |
| DIRMOBFOR | der-mob-for | Director of Mobility Forces |
| DME | D.M.E. | Distance Measuring Equipment |

| | | |
|---|---|---|
| DO | D.O. | Director of Operations |
| DV | D.V. | Distinguished Visitor |
| DZ | D.Z. | Drop Zone |
| ECP | E.C.P. | Entry Control Point |
| DIPS | dips | Diplomatic Clearances |
| ETP | E.T.P. | Equal Timing Point |
| FAST | fast | Fly-Away Security Team |
| FL | Flight Level | Flight Level |
| GPWS | jip-wizz | Ground Proximity Warning System |
| HVCDS | H.V.C.D.S. | High-Velocity Container Delivery System |
| IFF | I.F.F. | Identification of Friend or Foe |
| IFG | I.F.G. | In-Flight Guide |
| ILS | I.L.S. | Instrument Landing System |
| ISAF | eye-saff | International Security Assistance Force |
| LCLA | L.C.L.A. | Low-Cost Low Altitude aerial delivery system |
| LFA | L.F.A. | Legal for Alert |
| LFB | L.F.B. | Legal for Bravo |
| LOAC | low-ak | Law of Armed Conflict |
| LZ | L.Z. | Landing Zone |
| MEDEVAC | med-evac | Medical Evacuation |
| MPC | M.P.C. | Mission Planning Cell |
| NATO | Nay-tow | North Atlantic Treaty Organization |
| NVG | N.V.G. | Night Vision Goggles |
| PIFR | Piff-er | Pre-coordinated Instrument Flight Route |
| PL | Papa Lima | Positive Launch |
| PPR | P.P.R. | Prior Permissions Required |
| PT | P.T. | Physical Training |
| SCNS | skins | Self-Contained Navigational System |
| SDO | S.D.O. | Squadron Duty Officer |
| SKE | ski | Station Keeping Equipment |
| TAC/C2 | tac-see-two | Tactical Command and Control |
| TACAN | tac-can | Tactical Air Navigation |

| | | |
|---|---|---|
| TACC | T.A.C.C. | Tanker Airlift Control Center |
| TACP | Tac P | Tactical Air Control Party |
| TCAS | T-cass | Traffic Collision Avoidance System |
| TDD | T.D.D. | Theater Direct Delivery |
| TOLD | told | Take Off and Landing Data |
| UDM | U.D.M. | Unit Deployment Manager |
| UHF | U.H.F. or uniform | Ultra High Frequency (military used radios) |
| ULN | U.L.N. | Unit Line Number |
| VHF | V.H.F. or victor | Very High Frequency (civilian used radios) |
| VOR | V.O.R. | VHF Omnidirectional Range |
| VORTAC | vortac | VOR/TACAN |

Made in United States
Orlando, FL
04 November 2026

90201770R00157